Freshwater Fish Parasites

Freshwater Fish Parasites

Dr. Gadadhar Dash, M.F.Sc., Ph.D.
Associate Professor and Head,
Dept of Aquatic Animal Health
F.F.Sc.., W.B.U.A.F.S, Chakgaria Campus
Kolkata-700094, W.B., India

New India Publishing Agency
Pitam Pura, New Delhi-110 088

Published by

Sumit Pal Jain for

New India Publishing Agency

101, Vikas Surya Plaza, CU Block, L.S.C. Mkt.,
Pitam Pura, New Delhi-110 088, (India)
Phone : 011-27341616, Fax : 011-27341717
Mobile : 09717133558
E-mail : info@nipabooks.com
Web : www.nipabooks.com

ISBN: 978-93-81450-25-3

Print and Designed by NIPA

PREFACE

Diseases are the major constraints to sustainable development of Aquaculture industry. Many diseases affecting present day aquaculture is the resultant intensification of aquaculture practices without basic perception of intricate balance between host, pathogen and the environment. Many times disease out break in fish are linked to environmental detoriation and stress associated with intensification of culture practices. Stress factors such as inadequate physicochemical and microbial quality of culture water , poor nutritional status, high stocking density can cause infection by opportunistic pathogens. Among the commonly occurring pathogens, parasites causes more than 50% problems in the field and some times cause havoc among the fisherman community. Parasites when present in less numbers are not problematic, but when exceeds the limits cause severe problems, some times may lead to mass mortality.

The entitled book "Freshwater Fish Parasites" spread over 10 chapters, with complete pictorial guide of different fresh water fish parasites all over the country. The detail etiology, clinical signs , prophylaxis and treatment of parasitic diseases have been discussed in a simplified manner for easy understanding of students,researchers,teachers,scientists and farmers in the field .The book covers most interesting chapters and will guide to identify and manage parasitic diseases including some aspects of latest developments in fish parasitological research. A comprehensive approach has been made to cover all the progressive areas of parasitic disease management starting from identification of the parasites, basic status of diseases towards various approaches to diagnose, prevent and control disease conditions. The parasitic problems of different fishes, like indigenous fishes, exotic fishes of commercial importance, sport fishes and ornamental fishes have been addressed to give a complete parasitic disease profile on total fishery. The behavior of the hosts, interrelationship of soil and water with parasitized hosts, and mechanism of parasitic attachment are important aspects of the concerned book.

Date: 15.07.2011

Gadadhar Dash

CONTENTS

1

Parasites - An Overview

WHY, WHEN AND HOW FISH GET PARASITES

Parasites pose a real risk to fishes

All fish are potential hosts to many different species of parasites. Small numbers of parasites are common and probably do little harm. However, all parasites have tremendous reproductive potential and can, under ideal conditions, quickly overwhelm fish in the confines of a tank or pond.

Types of parasites

We can broadly class parasites into two groups. **Ectoparasites** are found on the external surfaces such as skin, fins and gills, while **endoparasites** are found in the internal tissues and organs.

Endoparasites are uncommon in ornamental fish. However, there is a wide diversity of blood parasites, worms that colonize the intestines and other parasites that can invade various organs and tissues. Many require an intermediate host, such as snails, birds or the introduction of an infected fish, and so are rare in pond/ aquarium fish. There are a variety of clinical signs which might indicate the presence of endoparasites. Lethargy accompanied by emaciation is a common sign, as are worms protruding from the anus. Identification of sporozoan and protozoan endoparasites often requires microscopic examination of tissues (Chanratchakool *et al.* 1994).

Ectoparasites are a common problem

For hobbyists, endoparasites are the most common parasitic problem. With few exceptions, these parasites are not life-threatening in small numbers, indeed small populations are fairly normal. The danger from endoparasites comes from their tremendous reproductive potential. In a natural environment this rarely leads to severe outbreaks, as only small numbers of juveniles would survive and find a host after hatching. Limited physical contact between fish would prevent parasites being transferred to a new host, so limiting the spread of the infestation although individual fish may be severely parasitised.

The confines of a pond or tank, with fish constantly coming into contact with each other, provide ideal conditions for fish-to-fish parasite transfer. The ready-availability of fish hosts results in high survival rates of juveniles hatched from eggs and cysts deposited in the pond. This can lead to serious parasite outbreaks affecting many fish (Conroy *et al.* 1997).

Fish can control numbers

The mucus layer, or cuticle, provides fish with some protection from parasites. In healthy fish it is continuously sloughed off, making it difficult for parasites to get a good foothold. Additionally, it contains various substances such as lysozyme, specific immunoglobins and free fatty acids, which are believed to have anti-pathogen properties. I stress that this assumes a healthy fish, living in a stress-free environment.

The situation changes when the fish are stressed, or the cuticle is being affected by adverse water quality such high ammonia/nitrite levels, inappropriate pH or other toxins. Under such conditions the consistency and protective properties of the cuticle can be severely altered, allowing parasites to proliferate.

Cuticle doesn't protect against all parasites

Although the cuticle offers some protection against parasite infestation, there are several parasites that are not hindered by its presence. These are mainly the larger parasites such as *Argulus* (fish louse) and *Lernaea* (anchor worms), but a very important exception is *Ichthyophthirius* (white-spot).

Secondary infections

In large numbers ectoparasites can cause severe damage to the integument from their feeding activities and /or their constant movement and attachment structures such as hooks and suckers. This in turn can often lead to secondary bacterial infections, typically body ulcers and bacterial gill disease.

The irritation caused by ectoparasites often leads to excess mucus production, seen as a grey, slime film, and epithelial hyperplasia, causing respiratory problems if the gills are affected.

Signs of parasite infestation

The larger parasites such as leeches, lice and anchor worms are visible with the naked eye. Typical signs of parasite infestations are rubbing, 'flashing' [there is a flash of silver from the underside of the fish it turns its body to rub against something], focal redness and inflammation on the body - often at the base of the dorsal fin, flared operculum, respiratory difficulties, lethargy and bacterial ulcers. However, it is important to realise that these 'symptoms' are not exclusive to parasites and may be caused by poor water quality or other diseases.

PARASITISM

Parasitism may be defined as an association of two nonspecific organisms of which one lives at the expense other.

The organism which lives at the expense of the other for food and shelter is called a parasite and one who is the victim is known as host.

A host is capable of living without the parasite but the latter can not live without a host.

Although a parasite lives at the cost of the host but it's aim is not to cause fatal damage to the host in order obtain a regular supply of food.

Types of Parasitism

Parasitism is a specialized type of life and the parasites have adopted various modifications and adaptations to lead this life successfully. The followings are different types of parasitism.

1) Accidental parasitism :-

   In this type, free living animal, when taken accidentally by the host, is capable of leading a parasitic life a certain period.

   Ex: Mud living nematodes.

2) Facultative Parasitism:-

   This is the type in which an animal can live both as parasitic and free living beings . Ex-mosquito.

3) Obligatory Parasitism:-

   In this type, a parasite has lost all power of leading a free life and most in or on some other organism during all or part of its life and it dies if perverted from doing so, Ex. Hook-warm

4) Latent parasitism – Which has no obvious effects on the host.

Kinds of Parasites

The followings are the different kinds of parasites according to their association with their host.

1) Intermittent Parasites:-

   They are occasional parasites which visit and leave their hosts at intervals.

Ex : Mosquito, Flea, bedbug.

2) Temporary Parasites:-

A parasite lives in or on a host only for short periods.They are parasitic on their host for some time and then lead a free life, Ex- leeches.

3) Permanent Parasites :-

A parasite which is associated with the host for a prolonged period. These parasites live permanently on the body of their host and can not live without them. Ex – Taenia, Fasciola.

Sometimes the parasites are classified as to weather, They are external or internal parasites. Followings are different kinds of parasites according to their habit and habitat.

1) Ecotozoic parasites :–

These are external parasites which are found on the body of their host. They suck the blood, feed upon, hour, feathers, skin or secretions of skin, ex-lice, Mosquitoes, ficks.

2) Entozoic parasites :-

These are internal parasites which live in the alimentary canal of their host.

Ex. Taenia, Ascaris.

3) Coelozoic parasites :-

These are the internal parasites which live in the coebnic cavity of their hosts.

Ex.: Microcystis, Taenia

Coebess : Main body cavity in which the gut is suspected.

4) Histozoic parasites : They live in the cells or tissues of their hosts. Ex-Trypanosoma, Plasmodium.

The parasites are placed into two categories according to whether they do considerable harm are not to their hosts.

1) Pathogenic parasites:-

They are also called imperfect parasites. They cause much harm to their hosts by destroying their tissues either temporarily or permanently. They may also produce toxic substances which are harmful or may be prove fatal to their hosts.

2) Non pathogenic parasites:-

They are known as perfect parasites. They cause little or no harm to their hosts.

The hosts are also classified as.

i) Definitive host / Primary host / Final host:

In this host the parasite reaches sexual maturity.

ii) Intermediate host/ Secondary host / Alternative host.

In this host the parasite passes preliminary stages of development.

iii) Obligatory host/Mandatory host/Required host :

The host species which is absolutely essential for existence of a parasite.

iv) Temporary host :

It is a host in or an which the parasite lives briefly then leaves to become free living.

v) Paratenic host : A host in which there is no development of immature parasite.

vi) Transport host : A host that carries undeveloping parasite to another host.

vii) Index parasite – Parasite indicating some ancient conditions.

viii) Intraspecific parasite – An organism parasitizing on organism of the same sps.

ix) Interspecific parasite – An organism parasitizing on organism of the different sps.

x) Satelite parasite – Species found in few host individuals and at low infection intensities.

xi) Hyper parasite – is a parasite on or in another parasite and there may be parasites of first, 2nd , 3rd or 4th degree.

xii) Microparasite – A parasite which resproduess directly and usually at high rates in or on the host, usually small with short generation time, inducing immunity and of transient nature. Ex. Protozoans, Bacteria, Virus

xii) Macroparasite – A parasite without direct reproduction is or on the host, usually large in size with long generation time, inducing immune response & that usually depends on infection and is of short duration.

Ex – Most helminths, Arthopods

Parasitic Castration – Sterility caused by the parasite.

Phoresis – One organism uses another as a means of transport without establishing a close association..

Preventing and treating fish parasites

Knowledge is power, and this is true even in the field of preventing and treating fish parasites. As a fish keeper, it is always a good idea to familiarize yourself with the various common fish parasites that you might encounter and learn more about how to detect them in your aquarium. Parasite infestations that are observed at an early stage is usually much easier to cure than parasites infestations that the aquarist does not notice until all the fishes are severely infested. By learning more about fish parasites and how to recognize them you can also decrease the risk of bringing home infected fish from the fish shop.

A lot of factor will play a role when it comes to determining the best cause of action when one suspect parasites in the aquarium. The species of parasite is naturally of imperative importance, but the fish species in the aquarium must also be taken into consideration since some fish species are exceedingly sensitive to certain parasite medications.

You must also decide if you want to treat the entire aquarium, or if it is better to move the fish to a separate hospital aquarium. In some situations, dipping the fish in a bowl of medication will be sufficient. The volume of your aquarium is also a factor to take into account since some medications are very expensive. (Sometimes euthanizing and replacing all the fish will be cheaper than trying to cure them.)

Generally speaking, aquarists treat their fish at two different times. The first time is before introducing a new fish to the aquarium, and the second time is when a possible parasite infection has been detected. When it comes to preventative treatment, the treatment will normally take place in a separate quarantine aquarium in order to avoid treating healthy fish. Before starting the treatment, always ask yourself if it is really necessary or if simply keeping the new fish in the quarantine aquarium and looking out for any symptoms of ill health would be enough to safe-guard against parasites.

In some situations, it will also be a good idea to move a wounded fish to a quarantine aquarium since a wounded fish can serve as a breeding ground for commonly occurring parasites that would normally not be able to affect the healthy fish in your aquarium. You can have parasites lurking around in your aquarium for years without noticing it, until a wounded fish gives them a chance to multiply and start attacking healthy fish in huge numbers. As you can see, the use of a quarantine aquarium can make parasite treatment unnecessary in many situations. This is a good thing since most parasite treatments are stressful for the fish (Das *et al* 1986).

Manual removal of external parasites and gill parasites

Some parasites that infest the skin and/or gills of the fish can be removed using forceps. Argulus, leeches, copepods, and isopods are all examples of such parasites.

Start by anesthetize your fish using Tricaine Methanesulfonate anesthetic (commonly sold under the name MS-222). Slowly add Tricaine Methanesulfonate to a minimal amount of water with the right salinity to anesthetize the fish. Use forceps to remove external/gill parasites that are large enough. If your fish is infested with gill isopods you will have to cut the legs of the isopods using small scissors and then use forceps to gently pull the legs from the gills.

Keep a watchful eye on the gills to make sure they continue to work while you work on your fish.

Make the process short and efficient, a fish that is kept in Tricaine Methanesulfonate for any longer period of time will die.

Dipping away external parasites

Generally speaking, saltwater dips will remove a lot of external parasites from freshwater species and freshwater dips will do the same for saltwater species. The exact length and frequency of the dips will vary depending on parasite species and fish species. In many cases, three dips that last for five minutes each will be enough. Dipping is stressful for the fish and must be used with caution.

Anti-parasite medication

In order to pick the right medication, you must first identify the culprit. There is no quick fix that will work for all possible fish parasites in the world, even though there are some remedies that are effective against a long row of parasites.

Always double-check the recommendations from the manufacturer of the medication by doing an online search. Some manufacturers want you to use a little too much medication. There is often a fine line between killing the parasites and killing the fish. Experienced aquarists can help you out with the right dosing. Also keep in mind that your particular fish species might be especially sensitive to the active ingredient (Gopalakrishnan, 1963)

A not about filters

Keep in mind that a lot of medications capable of killing microorganisms will kill off the beneficial bacteria in your aquarium. Take appropriate measures to avoid spikes in organic waste and rebuild your biological filter as soon as possible. An increase of organic waste is very dangerous for fish that are already weakened by parasites and a tough treatment.

❑❑❑

2

Introduction to Freshwater Fish Parasites

INTRODUCTION

Fish culture as a hobby or business is well established in India. Increased interest in fish culture has also increased awareness and experience with parasites that affect fish health, growth, and survival. Information provided in this chapter is intended for the novice fish culturist as a guide to common parasites of freshwater fish. Identification of parasites and their basic treatment is included; however, any herbal treatment alone or biological cum chemical treatment may be choosen for eradication purpose (Das et al., 1995a).

I. Identification of a Parasitic Problem

II. Protozoa

- A. Ciliates
  - 1. *Ichthyophthirius multifiliis*
  - 2. *Chilodonella*
  - 3. *Tetrahymena*
  - 4. *Trichodina*
  - 5. *Ambiphyra*
  - 6. *Apiosoma*
  - 7. *Epistylis*
  - 8. *Capriniana*
  - 9. Treatment of Ciliated Protozoan Infections
- B. Flagellates

IDENTIFICATION OF A PARASITIC PROBLEM

A common mistake of fish culturists is misdiagnosing disease problems and treating their sick fish with the wrong medication or chemical. When the chemical doesn't work, they will try another, then another. Selecting the wrong treatment because of misdiagnosis is a waste of time and money and may be more detrimental to the fish than no treatment at all. The majority of fish parasites can only be identified by the use of a microscope. If a microscope is unavailable, or the person using it has no previous experience with one, the diagnosis is difficult and questionable. Successful fish culturists learn by experience. Newcomers to the field need to learn the fundamentals of diagnostic procedures and how to use a microscope to identify parasites by attending short training courses. The following descriptions of common parasites can be used as references for understanding a professional diagnostic report or as a quick reference for the experienced fish culturist(Gopalakrishna,1964).

PROTOZOA

Most of the commonly encountered fish parasites are protozoans. With practice, these can be among the easiest to identify, and are usually among the easiest to control. Protozoans are single-celled organisms, many of which are free-living in the aquatic environment. Typically, no intermediate host is required for the parasite to reproduce (direct life cycle). Consequently, they can build up to very high numbers when fish are crowded causing weight loss, debilitation, and mortality. Five groups of protozoans are described in this publication: ciliates, flagellates, myxozoans, microsporidians, and coccidians. Parasitic protozoans in the latter three groups can be difficult or impossible to control as discussed below (Roberts, 2001).

Ciliates

Most of the protozoans identified by aquarists will be ciliates. These organisms have tiny hair-like structures called cilia that are used for locomotion and/or feeding. Ciliates have a direct life cycle and many are common inhabitants of pond-reared fish. Most species do not seem to bother host fish until numbers become excessive. In aquaria, which are usually closed systems, ciliates should be eliminated. Uncontrollable or recurrent infestations with ciliated protozoans are indicative of a husbandry problem. Many of the parasites proliferate in organic debris accumulated in the bottom of a tank or vat. Ciliates are easily transmitted from tank to tank by nets, hoses, or caretakers' wet hands. Symptoms typical of ciliates include skin and gill irritation displayed by flashing, rubbing, and rapid breathing.

Ichthyophthirius multifiliis

The disease called "Ich" or "white spot disease" has been a problem to aquarists for generations. Fish infected with this organism typically develop small blister-like raised lesions along the body wall and/or fins. If the infection is restricted to the gills, no white spots will be seen. The gills will appear swollen and be covered with thick mucus. Identification of the parasite on the gills, skin, and/or fins is necessary to conclude that fish has an "ich" infection. The mature parasite (Figure 1) is very large, up to 1000 µm in diameter, is very dark in color due to the thick cilia covering the entire cell, and moves with an amoeboid motion. Classically, *I . multifiliis* is identified by its large horseshoe-shaped macronucleus. This feature is not always readily visible, however, and should not be the sole criterion for identification. Immature forms of *I . multifiliis* are smaller and more translucent in appearance. Some individuals have suggested that the immature forms of *I . multifiliis* resemble *Tetrahymena* . Fortunately, scanning the preparation will usually reveal the presence of mature parasites and allow confirmation of the diagnosis.

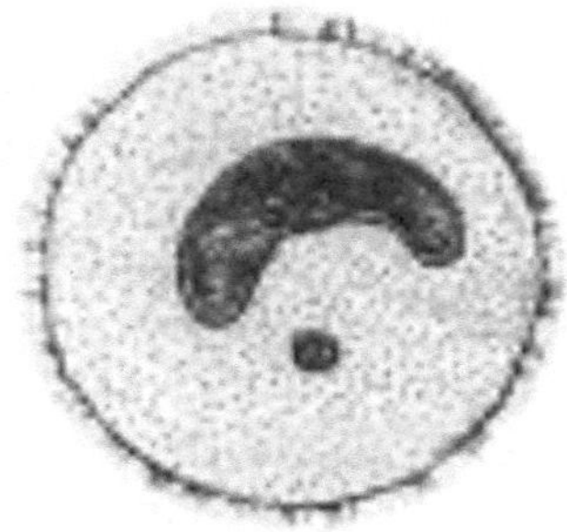

Figure 1.

If only one parasite is seen, the entire system should be treated immediately. "Ich" is an obligate parasite and capable of causing massive mortality within a short time. Because the encysted stage (Figure 2) is resistant to chemicals, a single treatment is not sufficient to treat "Ich". Repeating the selected treatment (Table- 1 & 2) every other day (at water temperatures 68-77°F) for three to five treatments will disrupt the life cycle and control the outbreak. Daily cleaning of the tank or vat helps to remove encysted forms from the environment (Paria, *et al.*1999).

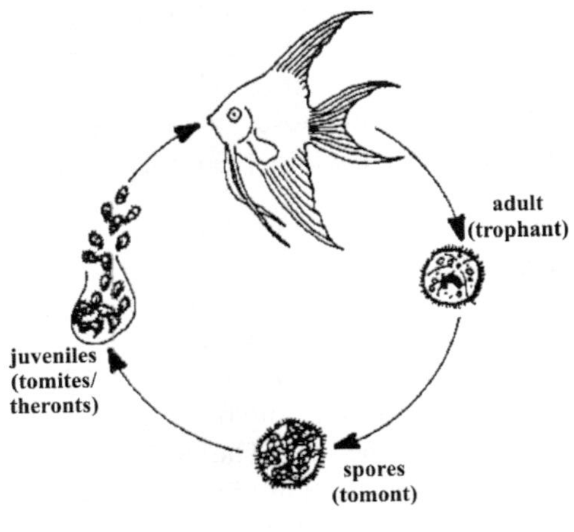

Figure 2.

Chilodonella

Chilodonella is a ciliated protozoan that causes infected fish to secrete excessive mucus. Infected fish may flash and show similar signs of irritation. Many fish die when infestations become moderate (five to nine organisms per low power field on the microscope) to heavy (greater than ten organisms per low power field). *Chilodonella* is easily identified using a light microscope to examine scrapings of skin mucus or gill filaments. It is a large, heart-shaped ciliate (60 to 80 m) with bands of cilia along the long axis of the organism (Figure 3). The organism is easily recognized at 100X magnification. *Chilodonella* can be controlled with any of the chemicals listed in Table- 1& 2 , and one treatment is usually adequate. *Chilodonella* has been eliminated in tanks using recirculating water systems by maintaining 0.02% salt solution (Das *et al.* 1997).

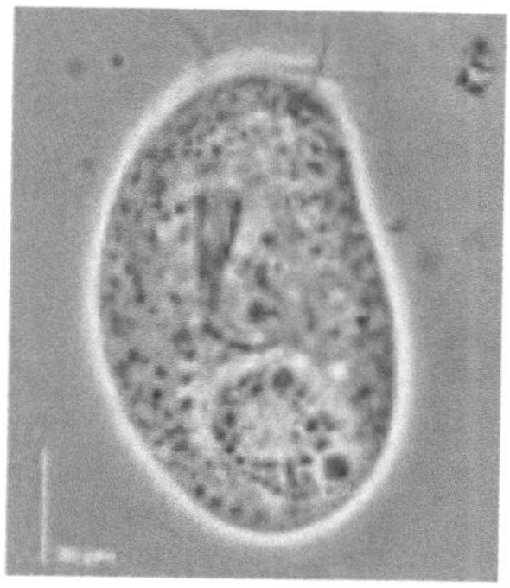

Figure 3.

Tetrahymena

Tetrahymena is a protozoan commonly found living in organic debris at the bottom of an aquarium or vat. *Tetrahymena* is a teardrop-shaped ciliate (Figure 4) that moves along the outside of the host. The presence of *Tetrahymena* on the body surface in low numbers (less than five organisms per low power field) is probably not significant. It is commonly found on dead material and is associated with high organic loads. Therefore, observing *Tetrahymena* on fish, which have been on the tank bottom, does not imply the parasite is the primary cause of death. One treatment of a chemical listed in Table- 1&2 should be adequate for control.

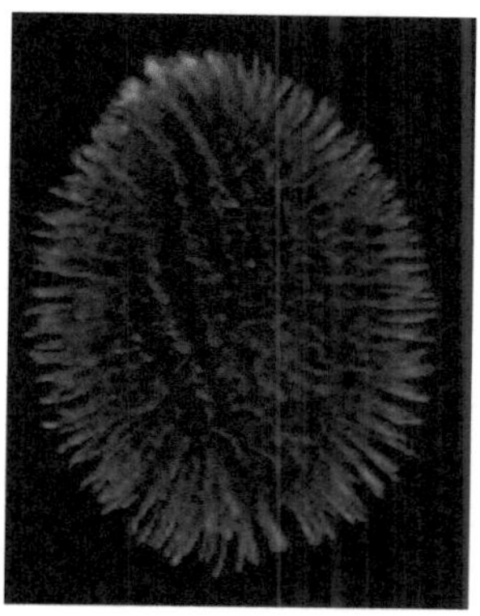

Figuare 4.

Table 1.

Chemical treatments for the control of external ciliates. "X" indicates that the chemical should not be used for this type of treatment.

Chemical Immersion	Dip	Short-term Bath	Prolonged (indefinite)
Copper sulfate	X	X	total alkalinity/100 (up to 2.5 mg/L), Do not use if total alkalinity < 50mg/L
Potassium permanganate	X	10 mg/L, 30 min	2 mg/L
Formalin	X	150—250 mg/L, 30 min	15—25 mg/L (2 drops/ gallon or 1 mL/10 gallons)
Salt	3%, Duration is species dependent.	1%, 30 min to 1 hr, species dependent	0.02—0.2%

Table-2:

Freshwater Parasite Treatments

EXTERNAL PARASITES	TREATMENTS
Ich *Ichthyophthirius multifiliis & Trichodina & Ichthyobodo &Tetrahymena*	Salt, formalin, malachite green, other commercial products, inc. temp.
Gill flukes, skin flukes	Copper, malachite green, formalin, others
External protozoa, flukes	Malachite green and formalin, others
Velvet (dinoflagellates)	Antiparasite medication, salt, copper, inc. temp
Fish lice, gill maggots	formalin, potassium permanganate, metriphonate, other commercial prod.
Anchorworms	organophosphate, manual removal and then dab with antiseptic
Larval stage of digenetic fluke parasites	organophosphate, other commercial products
INTERNAL PARASITES	**TREATMENTS**
Blood parasites	If flukes: praziquantel, others difficult to treat
Nematodes (roundworms)	Use of appropriate anthelmintic drug, eliminate IM host, remove dead fish, piperazine
Spiny-headed worm	Appropriate anthelmintic
Tapeworms	Praziquantel
Digenetic flukes	Difficult to treat, praziquantel
Hexamita & Spironucleus	Metronidazole, other commercial products

Source: Dash, *et al.* 2008, Smith, 1972. Srivastava, *et al.* 1994 and Kabata,1985.

Identification of *Tetrahymena* internally is a significant but untreatable problem. A common site of internal infection is the eye. Affected fish will have one or both eyes markedly enlarged (exophthalmia). Squash preparations made from fresh material reveal large numbers (10 per low power field) of *Tetrahymena* associated with fluids in the eye. Fish infected with *Tetrahymena* internally should be removed from the collection and destroyed.

Trichodina

Trichodina is one of the most common ciliates present on the skin and gills of pond-reared fish. Low numbers (less than five organisms per low power field) are not harmful, but when fish are crowded or stressed, and water quality deteriorates, the parasite multiplies rapidly and causes serious damage. Typically, heavily infested fish do not eat well and lose condition. Weakened fish become susceptible to opportunistic bacterial pathogens in the water. *Trichodina* can be observed on scrapings of skin mucus, fin, or on gill filaments. Its erratic darting movement and the presence of a circular, toothed disc within its body (Figure 5) easily identify it. *Trichodina* can be controlled with any of the treatments from Table-1&2. One application should be sufficient. Correction of environmental problems is necessary for complete control (Paperna, 1991).

Figure 5.

Ambiphyra

Ambiphyra , previously called *Scyphidia* , is a sedentary ciliate that is found on the skin, fins, or gills of host fish. Its cylindrical shape, row of oral cilia, and middle bank of cilia identify *Ambiphyra* (Figure 6). It is common on pond-reared fish, and when present in low numbers (less than five organisms per low power field), it is not a problem. High organic loads and deterioration of water quality are often associated with heavy, debilitating *Ambiphyra* infestations. This parasite can be controlled with one application of any of the treatments listed in Table-1&2.

Figure 6.

Apiosoma

Apiosoma , formerly known as *Glossatella* , is another sedentary ciliate common on pond-reared fish. *Apiosoma* can cause disease if their numbers become excessive. The organism can be found on gills, skin, or fins. The vase-like shape and oral cilia are characteristic (Figure 7). *Apiosoma* can be controlled with one application of one of the treatments from Table-1&2 .

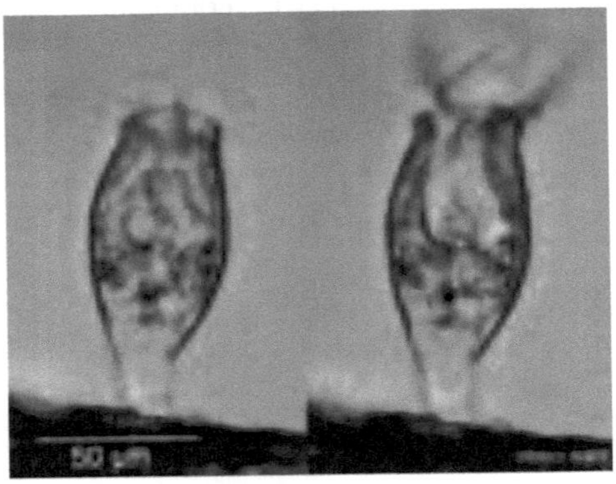

Figure 7.

Epistylis

Epistylis is a stalked ciliate that attaches to the skin or fins of the host. *Epistylis* is of greater concern than many of the ciliates because it is believed to secrete proteolytic (protein-eating) enzymes that create a wound, suitable for bacterial invasion, at the attachment site. It is similar in appearance to *Apiosoma* except for the non-contractile long stalk (Figure 8) and its ability to form colonies. In contrast to the other ciliates discussed above, the preferred treatment for *Epistylis* is salt. Fish can be placed into a 0.02% salt solution as an indefinite bath, or a 3% salt dip. More than one treatment may be required to control the problem (Mohan, *et al.* 1986). .

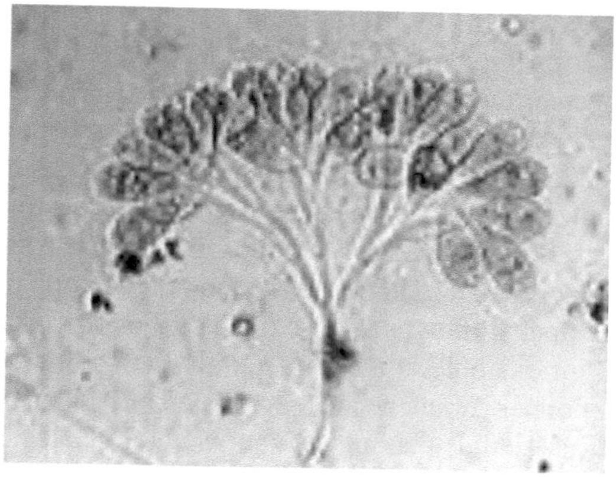

Figure 8.

Capriniana

Capriniana , historically called *Trichophyra* , is a sessile ciliate that attaches to the host's gills with a sucker. They have characteristic cilia attached to an amorphous-shaped body (Figure 9). In heavy infestations, *Capriniana* can cause respiratory distress in the host. One treatment from the chemicals listed in (Table 1& 2) should be adequate.

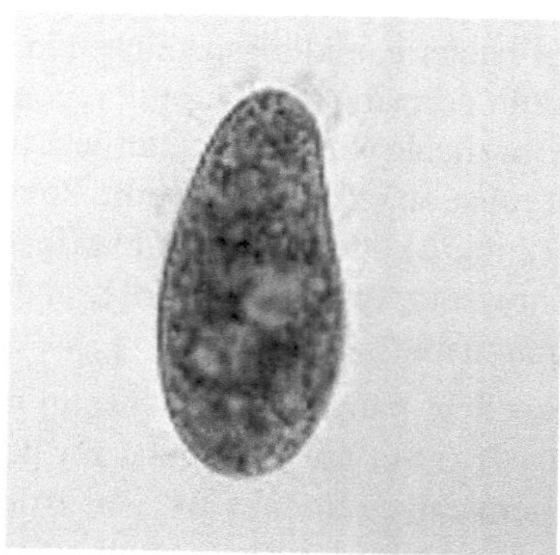

Figure 9.

Treatment of ciliated protozoan infections

Several chemicals commonly used to control ciliated protozoans in freshwater fish are listed below for your convenience. As stated above, most ciliate infestations respond to one chemical treatment; however, fish that do not improve as expected should be rechecked and retreated if necessary. Overtreatment with chemicals can cause serious damage to fish. Copper sulfate is an excellent compound for use in ponds to control external parasites and algae; however, it is extremely toxic to fish. Its killing action is directly proportional to the concentration of copper ions (Cu^{++}) in the water. As the alkalinity of the water increases, the concentration of copper ions in solution decreases. Consequently, a therapeutic level of copper in water of high alkalinity would be lethal to fish in water of low alkalinity. Conversely, a therapeutic concentration of copper in water of low alkalinity would be insufficient to have the desired action in water of higher alkalinity. For this reason, **the alkalinity of the water to be treated must be known** in order to determine the amount of copper sulfate needed. The amount of copper sulfate needed in mg/L is the total alkalinity (in mg/L) divided by 100. For example, if the total alkalinity in a pond is 100 mg/L, the concentration of copper sulfate needed would be 100/100 or 1 mg/L. If you are unsure how to measure the alkalinity of your water, or have never used copper sulfate, contact your aquaculture extension specialist for assistance. Never use copper sulfate in water that has a total alkalinity less than 50 mg/L.

Because of its algicidal activity, copper sulfate can cause dangerous oxygen depletions, particularly in warm weather. Emergency aeration should always be available when copper sulfate is applied to your system or ponds. Copper sulfate should not be run through the biofilter on a recirculation system as it will kill the nitrifying bacteria. If possible, tanks should be taken "off-line" during treatment with copper sulfate. If necessary, clean the biofilter manually to decrease organic debris and residual parasite load.

Potassium permanganate is effective against ciliates as well as fungus and external columnaris bacteria, and it can be used in a pond or vat. Multiple treatments with potassium permanganate are not recommended as it can burn gills. Aeration should be available when potassium permanganate is used because it is an algicide and can cause an oxygen depletion. Potassium permanganate at the prescribed dosage (2 mg/L) does not seem to affect the nitrifying bacteria in a biological filter; however, ammonia, nitrite, and pH should be closely monitored following treatment..

Formalin is an excellent parasiticide for use in small volumes of water such as vats or aquaria. It is not recommended for pond use because it is a strong algicide and chemically removes oxygen from the water. Vigorous aeration should always be provided when formalin is used.

Used in proper amounts, salt effectively controls protozoans on the gills, skin, and fins of fish. This is an effective treatment for small volumes of water such as aquaria or tanks. Use in ponds as a treatment is generally not recommended due to the large amount of salt and high cost of treatment that would be needed to be effective. Salt should never be used on fish that navigate by electrical field such as knifefish and elephant nose fish.

When using any treatment for fish, a bioassay (a test to determine safe concentration) should be conducted on a few fish before large numbers of fish are exposed. Fish species can react differently to various concentrations of the chemical; therefore, fish undergoing treatment must be monitored closely for adverse reactions. If the fish negatively react to treatment, the chemical should be flushed immediately from the system, or the fish should be moved to fresh water (Mohan, *et al.* 1986).

Flagellates

Flagellated protozoans are small parasites that can infect fish externally and internally. They are characterized by one or more flagella that cause the parasite to move in a whip-like or jerky motion. Because of their small size, their movement, observed at 200 or 400x magnification under the microscope, usually identifies flagellates. Common flagellates that infest fish are given below (Ferguson,1989).

Hexamita / Spironucleus

Hexamita is a small (3 -18 m) intestinal parasite commonly found in the intestinal tract of freshwater fish (Figure 10). Sick fish are extremely thin and the abdomen may be distended. The intestines may contain a yellow mucoid (mucus-like) material. Recent taxonomic studies have labeled the intestinal flagellate of freshwater angelfish as *Spironucleus* . *Hexamita* or *Sprironucleus* can be diagnosed by making a squash preparation of the intestine and examining it at 200 or 400x magnification. The flagellates can be seen where the mucosa (intestinal lining) is broken. They move by spiraling and in heavy infestations, they will be too numerous to be overlooked.

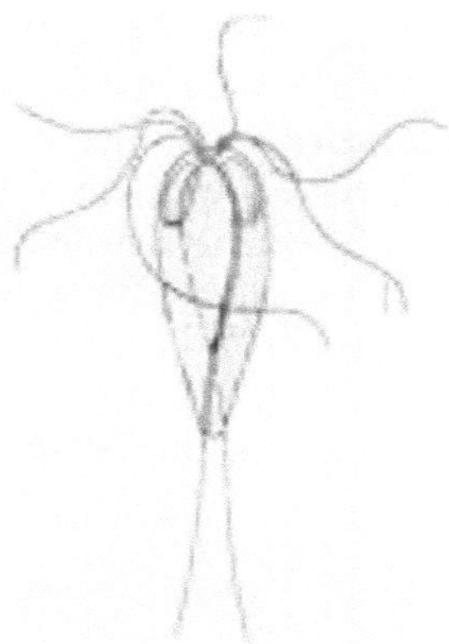

Figure 10.

The recommended treatment for *Hexamita / Spironucleus* is metronidazole (Flagyl). Metronidazole can be administered in a bath at a concentration of 5 mg/L (18.9 mg/gallon) every other day for three treatments. Medicated feed is even more effective at a dosage of 50 mg/kg body weight (or 10 mg/gm food) for five consecutive days.

Ichthyobodo

Ichthyobodo , formerly known as *Costia* , is a commonly encountered external flagellate (Figure 11). *Ichthyobodo*- infected fish secrete copious amounts of mucus. Mucus secretion is so heavy that catfish farmers popularly refer to the disease as "blue slime disease". Infected angelfish also produce excessive mucus that can give dark colored fish a gray or blue coloration along the dorsal body wall. Infected fish flash and lose condition, often characterized by a thin, unthrifty appearance. *Ichthyobodo* can be located on the gills, skin, and fins, however, it is difficult to identify because of its small size. The easiest way to identify *Ichthyobodo* is by its corkscrew swimming pattern. With a good microscope, the attached organism can be seen at 400x magnification. The organism is easily controlled using one application of one of the treatments listed in (Table 1&2).

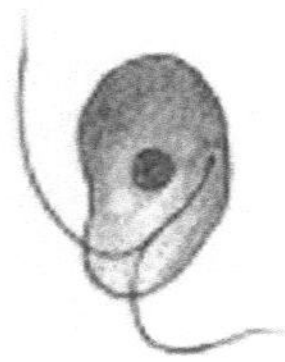

Figure 11.

Piscinoodinium

Piscinoodinium is a sedentary flagellate that attaches to the skin, fin, and gills of fish. The common name for *Piscinoodinium* infection is "Gold Dust" or "Velvet" Disease. The parasite has an amber pigment, visible on heavily infected fish. Affected fish will flash, go off feed, and die. *Piscinoodinium* is most pathogenic to young fish. The life cycle of this parasite can be completed in 10-14 days at 73-77°F (Figure 12), but lower temperatures can slow the life cycle. Also, the cyst stage is highly resistant to chemical treatment. Therefore, several applications of a treatment (Table 1&2) may be necessary to eliminate the parasite. For non-food species, chloroquin (10mg/L prolonged bath) has been reported to be efficacious.

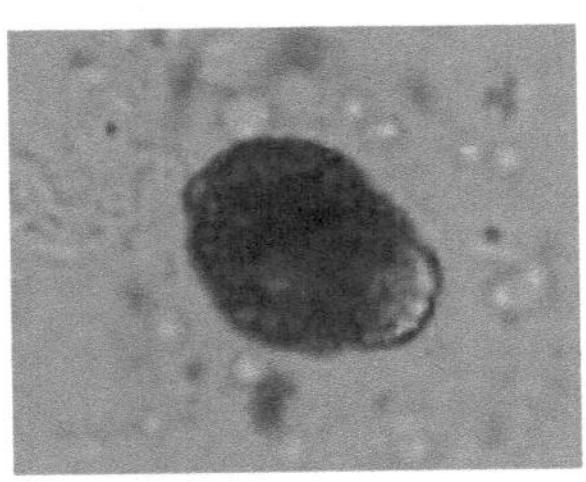

Figure 12.

Cryptobia

Cryptobia is a flagellated protozoan common in cichlids. They are often mistaken for *Hexamita* as they are similar in appearance. However, *Cryptobia* are more drop-shaped, with two flagella, one on each end. Also, *Cryptobia* "wiggles" in a dart-like manner, whereas *Hexamita* "spirals". *Cryptobia* typically is associated with granulomas (Figure 13), in which the fish "walls off" the parasite. These parasites have been observed primarily in the stomach, but may be present in other organs. Fish afflicted with *Cryptobia* may become thin, lethargic and develop a dark skin pigmentation. A variety of treatments are presently being studied with limited success. Nutritional management has proven to take an active role in its control.

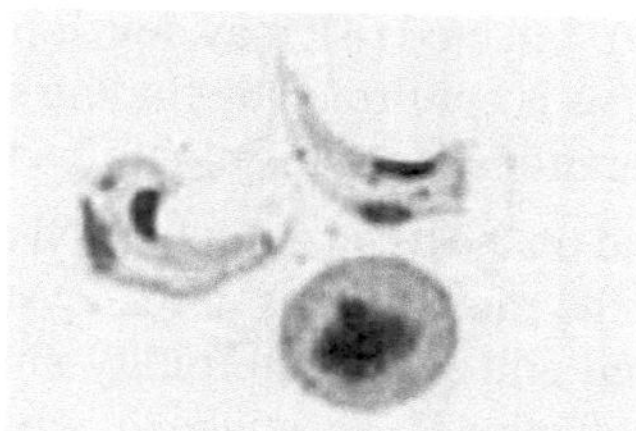

Figure 13.

Myxozoa

Myxozoa are parasites that are widely dispersed in native and pond-reared fish populations. Most infections in fish create minimal problems, but heavy infestations can become serious, especially in young fish. Myxozoans are parasites affecting a wide range of tissues. They are an extremely abundant and diverse group of organisms, speciated by spore shape and size. Spores can be observed in squash preparations of the affected area at 200 or 400x magnification or by histologic sections.

Clinical signs vary, depending on the target organ. For example, fish may have excess mucus production, observed with *Henneguya* (Figure 14) infections.

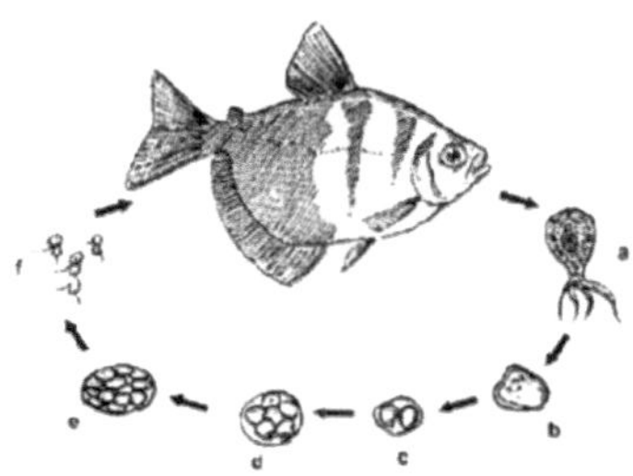

Figure 14.

White or yellowish nodules may appear on target organs. Chronic wasting disease is common among intestinal myxozoans such as with *Chloromyxum* . "Whirling disease" caused by *Myxobolus cerebralis* has been a serious problem in salmonid culture. Elimination of the affected fish and disinfection of the environment is the best control of myxozoans. There are no established remedies for fish. Spores can survive over a year, so disinfection is mandatory for eradication (Dash, et al. 2000).

Microsporidia

Microsporidians are intracellular parasites that require host tissue for reproduction. Fish acquire the parasite by ingesting infective spores from infected fish or food. Replication within spores (schizogony) causes enlargement of host cells (hypertrophy). Infected fish may develop small tumor-like masses in various tissues. Diagnosis is confirmed by finding spores in affected tissues, either in wet mount preparations, or in histologic sections.

Clinical signs depend on the tissue infected and can range from no visible lesions to mortalities. In the most serious cases, cysts enlarge to a point that organ function is impaired and severe morbidity and/or mortality results. A common microsporidian infection is *Pleistophora* , which infects skeletal muscle (Figure15).

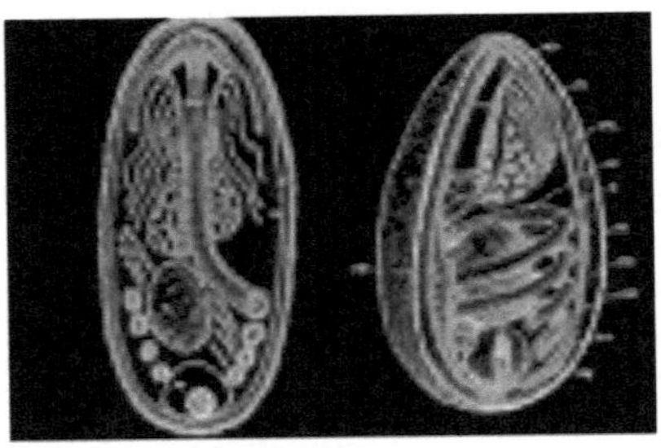

Figure 15.

There is no treatment for microsporidian infections in fish. Spores are highly resistant to environ-mental conditions and can survive for long periods. Elimination of the infected stock and disinfection of the environment is recommended (Roy, *et al.*2006)..

Coccidia

Coccidia are intracellular parasites described in a variety of wild caught and cultured fish (Figure 16). Their role in the disease process is poorly understood, but there is increasing evidence that they are potential pathogens. The most common species encountered in fish are intestinal infections. Inflammation and death of the tissue can occur, which can affect organ function. Other infection sites include reproductive organs, liver, spleen, and swim bladder.

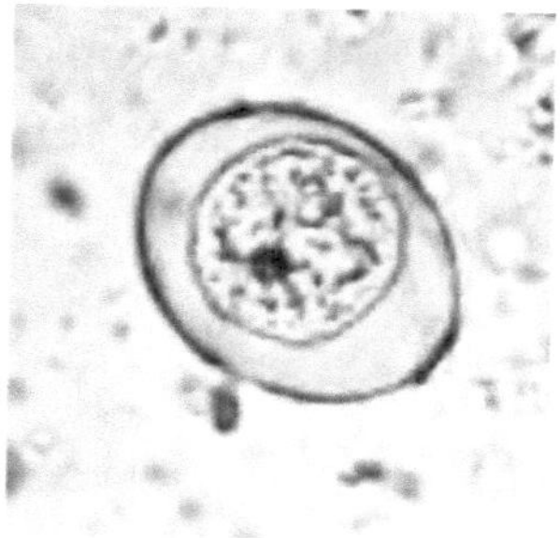

Figure 16.

Clinical signs depend on target organ affected but may include general malaise, poor reproductive capacity, and chronic weight loss. A definitive diagnosis of tissue coccidia should be completed with histologic or electron microscopy. Several compounds have been used to control coccidiosis with some success; however, consultation with an experienced fish health professional is recommended. Maintaining a proper environment and reducing stress appear to be important in preventing coccidia outbreaks in cultured fish.

MONOGENEAN TREMATODES

Monogenean trematodes, also called flatworms or flukes, commonly invade the gills, skin, and fins of fish. Monogeneans have a direct life cycle (no intermediate host) and are host- and site-specific. In fact, some adults will remain permanently attached to a single site on the host.

Freshwater fish infested with skin-inhabiting flukes become lethargic, swim near the surface, seek the sides of the pool or pond, and their appetite dwindles. They may be seen rubbing the bottom or sides of the holding facility (flashing). The skin where the flukes are attached shows areas of scale loss and may ooze a pinkish fluid. Gills may be swollen and pale, respiration rate may be increased, and fish will be less tolerant of low oxygen conditions. "Piping", gulping air at the water surface, may be observed in severe respiratory distress. Large numbers (>10 organisms per low power field) of monogeneans on either the skin or gills may result in significant damage and mortality. Secondary infection by bacteria and fungus is common on tissue with monogenean damage.

Gyrodactylus and *Dactylogyrus* are the two most common genera of monogeneans that infect freshwater fish (Figure 17 A & B). They differ in their reproductive strategies and their method of attachment to the host fish. *Gyrodactylus* have no eyespots, two pairs of anchor hooks, and are generally found on the skin and fins of fish. They are live bearers (viviparous) in which the adult parasite can be seen with a fully developed embryo inside the adult's reproductive tract. This reproductive strategy allows populations of *Gyrodactylus* to multiply quickly, particularly in closed systems where water exchange is minimal .

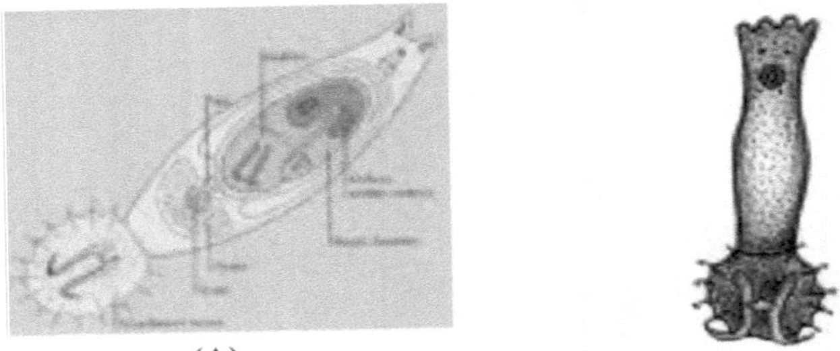

(A) (B)

Figure 17. A, B

Dactylogyrus prefers to attach to gills. They have two to four eyespots, one pair of large anchor hooks, and are egg layers. The eggs hatch into free-swimming larvae and are carried to a new host by water currents and their own ciliated movement. The eggs can be resilient to chemical treatment, and multiple applications of a treatment are usually recommended to control this group of organisms (Paperna, 1963b).

Treatment of monogeneans is usually not satisfactory unless the primary cause of increased fluke infestations is found and alleviated. The treatment of choice for freshwater fish is formalin, administered as a short-term or prolonged bath (Table 1&2). Fish that are sick do not tolerate formalin well, so they need to be carefully monitored during treatment. Potassium permanganate can also be effective in controlling monogeneans (Buchmann, *et al.*1987a.).

DIGENEAN TREMATODES

Digenean trematodes have a complex life cycle involving a series of hosts (Figure 18). Fish can be the primary or intermediate host depending on the digenean species. They are found externally or internally, in any organ. For the majority of digenean trematodes, pathogenicity to the host is limited.

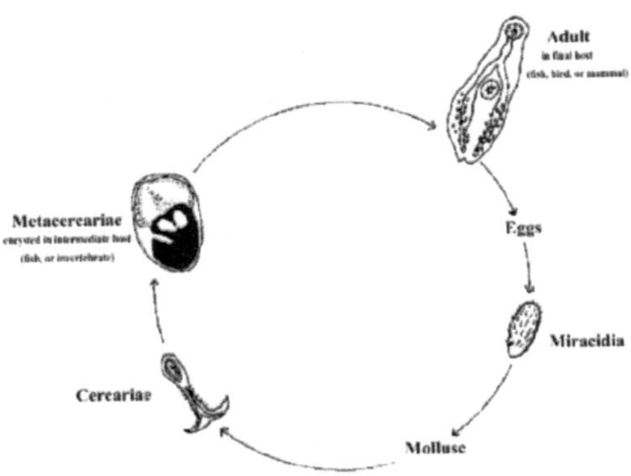

Figure 18.

The life stage most commonly observed in fish is the metacercaria, which encysts in fish tissues (Figure 19). Again, metacercaria that live in fish rarely cause major problems. However, in the ornamental fish industry, digenetic trematodes from the family Heterophyidae, have been responsible for substantial mortalities in pond-raised fish. These digeneans become encysted into gill tissue and respiratory distress is eminent.

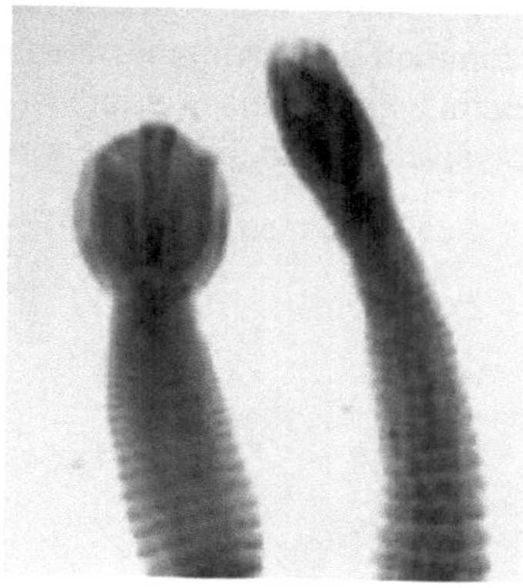

Figure 19.

Another example of a metacercaria that could cause problems in cultured fish is the genus *Posthodiplostonum* or the white grub. This has caused mortalities in baitfish, but usually the only negative effect is reduced growth rate, even when the infection rate is high. In cases where mortalities occur, there are unusually high numbers in the eye, head, and throughout the visceral organs.

Another fluke is *Clinostonum* , often called yellow grub. It is a large trematode and although it does not cause any major problems for fish, it is readily seen and will make fish unmarketable for aesthetic reasons.

The best control of digenean trematodes is to break the life cycle of the parasite. Elimination of the first intermediate host, the freshwater snail is often recommended. Copper sulfate in ponds has been used with limited success and is most effective against snails when applied at night, due to their nocturnal feeding activity. For Florida ornamental fish growers a chemical for snail control, Bayluscide, is available. This is a Restricted Use Pesticide and access to it is controlled. For more information, contact your aquaculture Extension specialist or the University of Florida Tropical Aquaculture Laboratory in Ruskin, Florida (Tweb, *et al.*1981).

NEMATODES

Nematodes, also called roundworms, occur worldwide in all animals. They can infect all organs of the host, causing loss of function of the damaged area.

Signs of nematodiasis include anemia, emaciation, unthriftiness and reduced vitality. Three common nematodes affecting fish are described (Dash, *et al.* 2004,2008).

Camallanus

Camallanus (Figure-20) is easily recognized as a small thread-like worm protruding from the anus of the fish. Control of this nematode in non-food fish is with fenbendazole, a common antihelminthic. Fenbendazole can be mixed with fish food (using gelatin as a binder) at a rate of 0.25% for treatment. It should be fed for three days, and repeated in three weeks.

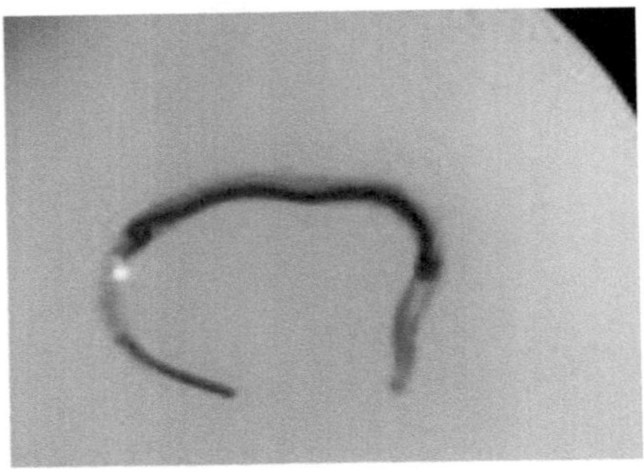

Figure 20.

Capillaria

Capillaria is a large roundworm commonly found in the gut of angelfish (Figure 21), often recognized by its double operculated eggs in the female worm (Figure 22). Heavy infestations are associated with debilitated fish, but a few worms per fish may be benign. Fenbendazole is recommended for treatment (see above) (Das, *et al.* 2006).

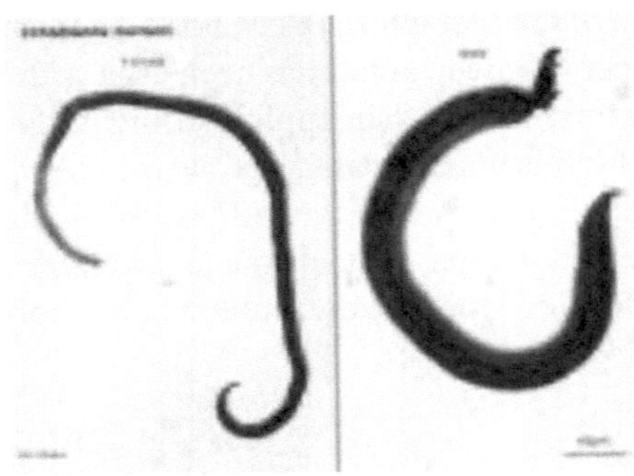

Figure 21.

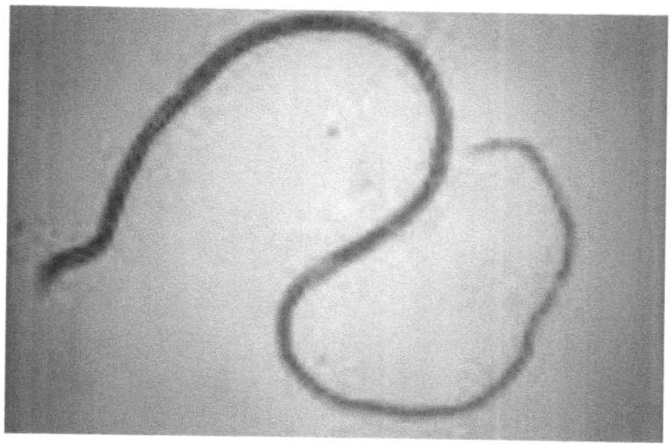

Figure 22.

Eustrongylides

Eustrongylides are nematodes that uses fish as its intermediate host. The definitive host is a wading bird, a common visitor to ponds. The worm encysts in the peritoneum or muscle of the fish and appears to cause little damage. Because of the large size of the worms (Figure 23), infected fish may appear unsuitable for retail sales. Protecting fish from wading birds and eliminating the intermediate host, the oligocheate or Tubifex (soft-bodied worms), are the best means to prevent infection.

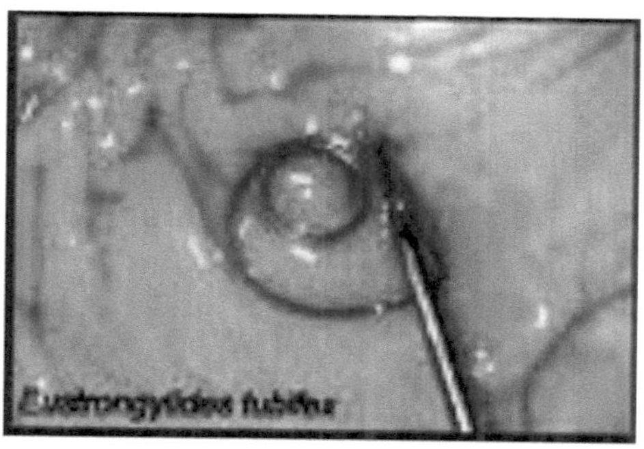

Figure 23.

CESTODES

Cestodes, also called tapeworms, are found in a wide variety of animals, including fish (Figure 24). The life cycle of cestodes is extremely varied with fish used as the primary or intermediate host. Cestodes infect the alimentary tract, muscle or other internal organs. Larval cestodes called plerocercoids are some of the most damaging parasites to freshwater fish. Plerocercoids decrease carcass value if present in muscle, and impair reproduc-tion when they infect gonadal tissue. Problems also occur when the cestode damages vital organs such as the brain, eye or heart (Dash, *et al.* 2004).

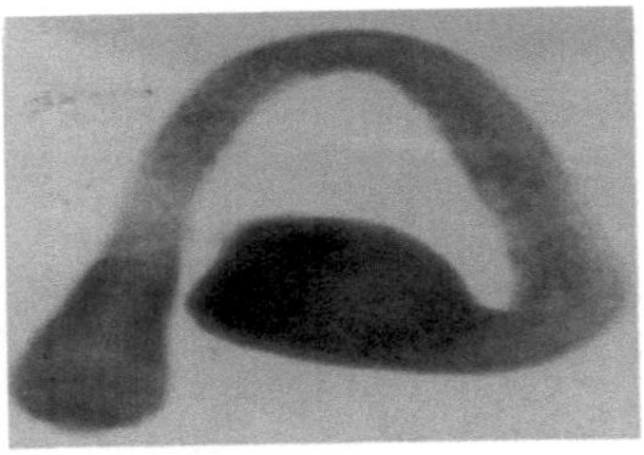

Figure 24.

One of the most serious adult cestodes that affect fish is the Asian tapeworm, *Bothriocephalus acheilognathi* . It has been introduced to the United States with grass carp and has caused serious problems with bait minnow producers.

Praziquantel at 2 -10 mg/L for 1 to 3 hours in a bath is effective in treating adult cestode infections in ornamental fish. At this time, there is no treatment that can be used for food fish. Also, there is no successful treatment for plerocercoids. Ponds can be disinfected to eradicate the intermediate host, the copepod.

PARASITIC CRUSTACEA

Parasitic crustacea are increasingly serious problems in cultured fish and can impact wild populations. Most parasitic crustacea of freshwater fish can be seen with the naked eye as they attach to the gills, body and fins of the host. Three major genera are discussed below (Yamaguti, 1963 and Overstreet, 1979).

Ergasilus

Ergasilus (Figure 25) are often incidental findings on wild or pond-raised fish and probably cause few problems in small numbers. However, their feeding activity causes severe focal damage and heavy infestations can be debilitating. Most affect the gills of freshwater fish, commonly seen in warm weather. A 3% salt dip, followed by 0.2 %-prolonged bath for three weeks, may be effective in eliminating this parasite.

Figure 25.

Lernaea

Lernaea , also known as anchor worm (Figure 26), is a common parasite of goldfish and koi, especially during the summer months. The copepod attaches to the fish, mates, and the male dies. The female then penetrates under the skin of the fish and differentiates into an adult. Heavy infections lead to debilitation and secondary bacterial or fungal infections. Removal of the parasite by hand with forceps may control lernaeid infestations with careful monitoring of the wound. A 3% salt dip followed by 0.2%-prolonged immersion has been used to effectively control *Lernaea* in goldfish and koi ponds.

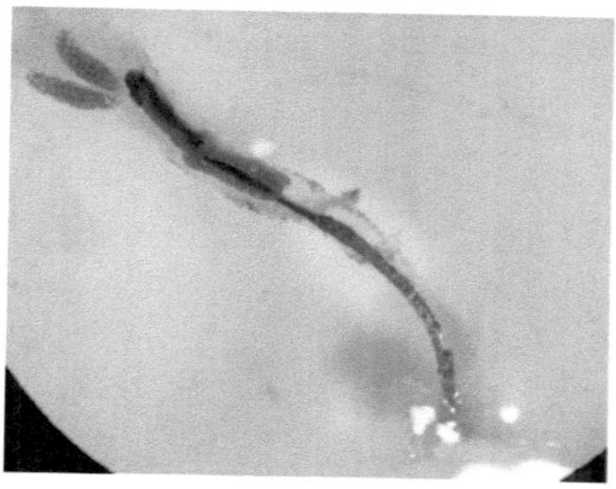

Figure 26.

Argulus

Argulus or fish louse is a large parasite (Figure 27) that attaches to the external surface of the host and can be easily seen with the unaided eye. *Argulus* is uncommon in freshwater aquarium fish but may occur if wild or pond-raised fish are introduced into the tank. It is especially common on goldfish and koi Individual parasites can be removed from fish with forceps, but this does not eliminate parasites in the environment. A prolonged immer-sion of 0.02 - 0.2% salt may control re-infection to the fish host (Parida, *et al.* 2008 and Hora,1943)

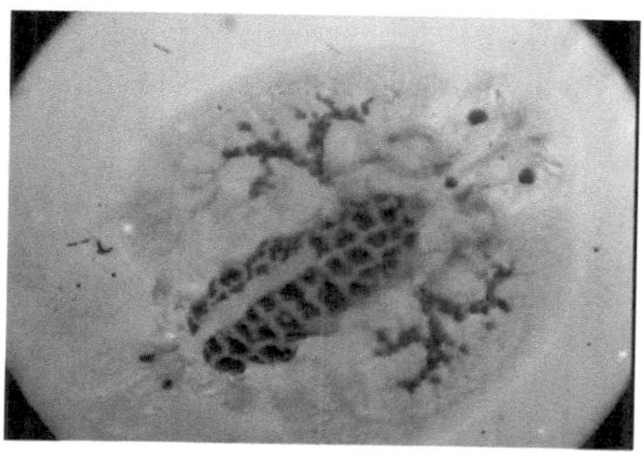

Figure 27.

LEECHES

Leeches are occasionally seen in wild and pond-raised fish. They have a direct life cycle with immature and mature worms being parasitic on host's blood. Pathogenesis varies with number and size of worms and duration of feeding. Heavily infested fish often have chronic anemia. Fish may develop secondary bacterial and fungal infections at the attachment site.

Leeches resemble trematodes but are much larger and have anterior and posterior suckers (Figure 28). Dips in 3% saltwater are effective in controlling leeches. Ponds with heavy leech infestation require drainage, treatment with chlorinated lime, followed by several weeks of drying. This will destroy the adults and their cocoons containing eggs (Roberts, 2001).

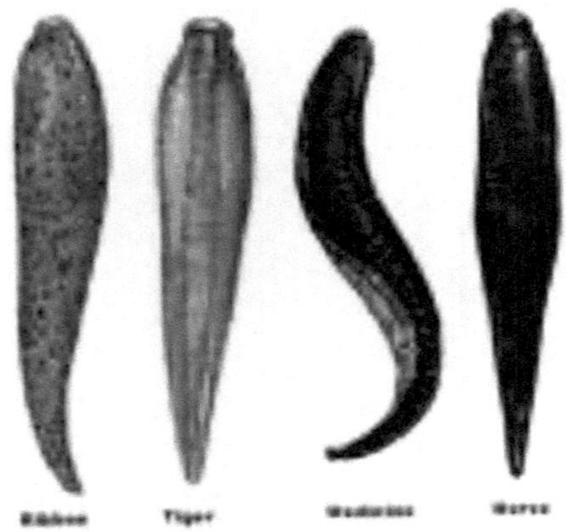

Figure 28.

CONCLUSION

Most fish health problems occur because of environmental problems: poor water quality, crowding, dietary deficiencies, or "stress". The best cure for any fish health problem is prevention. Good water quality management and proper fish husbandry techniques will eliminate most parasites described here.

3

Diseases, Parasites and other Maladies

PREVENTION

This section will discuss ailments that affect the fish kept in our aquariums. We will simply not call them diseases because most are not true diseases but a response to some sort of stimuli such as poor tank water management or the presence of actual parasites. A list of the most common ailments, the reasons behind them and the appropriate action needed to "fix" the problem is given in this Chapter.

If you maintain good water quality, keep the right kind of fish together, feed good quality food and monitor all your water properties you will rarely have any problems. A few common sense precautions when you purchase new fish can save you a lot of trouble.

A few things to observe:

1. Are the fish active?
2. Are they eating?
3. Are there any scars, spots or open wounds?
4. Are the gills nice and pink?
5. Are the fins wide open, not clamped close to the body?

All of these and more should be closely looked at before you buy any fish and never buy a fish that does not look or act right and "cure" it at home! One last important thing, check out the stores filtration system. If the system is a central one be sure to check out all the tanks for signs of parasites as many are free swimming at some point in their lives, and they could move freely from tank to tank (Mousavi, 2003).

Once you decide to purchase the fish a few more precautions should be taken in order to keep risk at a minimum.

1. Slowly acclimatize the fish to your tank water.
2. Use an antibiotic dip if you can.
3. Ideally have a quarantine tank set up.
4. Isolate the fish for two weeks in the quarantine tank.
5. Keep a close eye on the fish once it is in your tank.
6. Prevention is better than intervention.

These are basic common sense guidelines that can save your fish, plants and money if properly followed. If you find yourself confronted with one of these ailments or some other not described here, please follow for a comprehensive guide to some of the most popular medications available for our use below.

Diseases Table

Probably one of the most distressing times of your aquarium with you and your fish, will be when your fish are ill means, there you are, proud of your new fish and aquarium and - bam!! Your fish are covered in spots, or wool-like, small tufts on the body. Your mind will be racing, and thinking things like: What the hell is that stuff? Is it incurable? Will it infect all the other fish? Is this game over for all of my fish? Yes, its a worrying time, thats for certain. So to help you to catch an illness quickly, here are a few (only a few, there are hundreds of diseases around) of the more common tropical freshwater fish diseases.

Examples of fish diseases

To see an example of a Healthy Fish.

Symptons of illness	It could be...	Treatment
Your new fish have died - there are no obvious signs of illness	"New Tank Syndrome"	Adding fish to a brand new aquarium stresses the fish, and sometimes it can be fatal. For more info, check out: The Nitrogen Cycle
Loss of appetite	An enviromental problem, newly-introduced fish can be slow to feed (stress), certain fishes will refuse anything but live foods.	Enviromental: inspect for any manifestations of disease, and get a basic water tester (Ammonia etc). The same with stress. Be patient with them. Look for other signs of disease.
Cotton Wool effect on the body	Fungus, Columnarius (Mouth fungus)	If you see this disease, can use medication (ie "Anti Fungus") and a Salt bath (1 teaspoon per gallon)

Contd...

Small, white spots on the head, body and fins.	Whitespot (Ichthyophthirius)	A very common disease (especially in a new aquarium). There are several formulas around, dose the fish quickly before the parasite spreads. Salt could be used (1 teaspoon per galloon).
Fraying of the soft tissue (fin membrane)	Fin Rot, the disintegration of the fins	Use medication before a secondary infection occurs, it will take several days before there is serious fin damage. Could use salt bath (1 teaspoon per gallon)
Appearance of a whitish film over the eye of the fish	Poor water quality, poisoning (Chlorine) or Eye Fluke	Poor water: Water changes are required. Poisoning: Make sure to condition tap water. Eye Fluke: Use medication
Fish are gasping, on the bottom of the aquarium, red streaks around gills	Most probably oxygen starvation, or a Gill disease	You have too many fish, an air pump may help. Gill diseases are quite rare, you could actually see the gills rot away: use medication.
Worms are visibly seen hanging from the gills, fish are gasping	Gill Worm (Dactylogyrus)	This is another parasite. Use anti-parasite drugs as directed.
Reddining of the skin, or red streaks in the fins	Virus, Ulcers (Bacterial infection)	Viruses can only be seen through microscopes - and its incurable. With a bacterial disease, use a anti-bacteria or anti-internal treatment. A prolonged SALT bath (3-5 g/litre for several days) will alliviate osmoregulatory stress in the case of severe ulceration.
Scales seem to be raised, fish are bloated.	Dropsy (Ascites)	Dropsy is more of a symptom than an actually disease yet, and is quite common. Most associated with a bacterial infection (cause could be wide - poor water conditions are often the culprit. Meds can be used, but dont always work.
Small, worm like parasites attached to fish, fish scraping against rocks.	Anchor worm (lernea)	There are many Crustacean Parasites around, and can be exported into your water. Meds are available.

Contd...

Dull colour (mucus), detached skin, fraying of the fin.	Mucus Hyperproduction.	This called by caused by Slime disease (parasite, meds available), poor water conditioners (high Nitrogenous wastes), fFukes, stress.

A word of warning with using medication: Obviously people want to save their fish, so they go and buy a product, and sometimes these can be quite effective. But a word of warning - some meds can be very detrimental in the long term. Treatments which use Malachite Green and Methylene Blue can be useful treatments for Parasites and Fungus, unfortunately they can destroy our beneficial bacteria, and then you can get a Ammonia/Nitrite peak for some time. Also some meds demand it that they remove Carbon/Zeolite for the duration of the treatment - this can effectively start up a mini-cycle as the Carbon/Zeolite have lost their bacteria.

Salt bath: So a salt bath is a good way of treating the fish without damaging the bacteria we need. It is a good antiseptic and fungicide. If you are using high doses, its best to try to gradually raise the concentration, so the fish don't become stressed. Sometimes it can be effective to use a short salt bath, with higher salt concentration (such as fungal infection - 10G/Litre for 30 min).

DISEASE TREATMENT

Diseases

Under ideal circumstances - proper water conditions, a diet including a variety of foods, uncrowded conditions, and an environment lacking other stress-diseases rarely affect fish. Usually fish will only get sick when something in the environment is not right, or when new fish is added to the tank. Healthy fish generally have strong immune systems and are capable of resisting most pathogens, but when stress weakens the fish, the fish becomes more susceptible to illness. There are several circumstances, including non-living and living factors, which can lead to fish stress.

Non-living Sources of Stress

Among non-living factors that can adverse affects on aquarium (and terrestrial) inhabitants include poisoning, incorrect temperature or pH, and a shortage of oxygen.

Ammonia Poisoning

Ammonia poisoning is caused by the buildup of organic waste due to overfeeding, fish or plant deaths and decay, or improper cycling. Ammonia poisoning especially occurs when the pH exceeds 7, when benign ammonium becomes ammonia. Symptoms of ammonia poisoning include sluggish behavior, panting, and gill discoloration (gill burn). Fish may hang just below the water surface. The easiest way to confirm ammonia poisoning is by testing the water. Ammonia poisoning can be reduced by reducing feedings, making water changes, lowering the pH, using zeolites, and increasing aeration.

Nitrite/Nitrate Poisoning

Nitrite/Nitrate poisoning is caused by the same activities as ammonia poisoning. Nitrite/Nitrate poisoning has the same symptoms as ammonia poisoning and can be tested by a Nitrite/Nitrate water test kit. The best course of action, is to reduce feeding, make frequent partial water changes until the compounds are reduced, and increase the aeration in the water.

Chlorine Poisoning

Free chlorine, present in most tap water, is toxic to fish. Chlorine affects the gills and causes death by asphyxiation. Chlorine can be removed by boiling the water, letting the water stand for a few days, vigorously aerating the water, or by adding a water conditioner.

Heavy Metal Poisoning

Heavy metal poisoning can result from old pipes and/or metal in the fish tank. Heavy metal poisoning is evident when fish gasp at the surface for air and breath rapidly. Tests are available to measure the amounts of heavy metals in your water. The best way to remove heavy metals is to utilize a reverse osmosis system, although filtering the water through activated carbon and using water conditioners can be substituted.

Hydrogen Sulfide Poisoning

This gas is caused by rotting debris and waste in the gravel of the tank. This gas is toxic and the first signs of its presence are a rotten egg-like odor and fish gasping at the water surface for air. The best measure to take is to make a large water change, using a siphon to remove waste from the gravels. Make partial water changes until the odor is gone and the fish return to normal swimming and breathing.

Medication Poisoning

Medications are meant to help fish recover, although when misused, can be harmful than helpful. Medications can have adverse affects on many types of fish including catfish, tetras, Mormyrids, Loaches, and other sensitive fish. Copper-based medications have harmful affects on invertebrates, so always remove snails and crustaceans from the tanks before treating it. Always be sure to read the label on the medication to confirm that it is suitable for your fish. If a medication appears to be harming your fish, make a partial water change and filter the water with activated carbon.

Other Pollutant poising

There are other chemicals (cigarette smoke, paint fumes, pesticides) that sometimes make their way into the fish tank. The best way to combat these pollutants is not to allow them to get in the tank in the first place. However, once a foreign pollutant enters the tank, the results can be drastic for the inhabitants. Try making water changes and filtering with activated carbon to alleviate the problem.

Incorrect Temperature

Most tropical fish are used to living in water with a small temperature variance. When the temperature drops below or exceeds this range, fish can be weakened and left more vulnerable to disease. The best way to prevent wide variances in temperature is to purchase a reliable heater and place the tank away from drafty areas.

Incorrect pH

Most tropical fish live in water with a relatively stable pH. When the pH is not right, the fish are weakened and become more susceptible to illness and infection. Symptoms of an improper pH include darting movements, inflamed and bleeding gills, rapid gill movements, and fish hanging just below the water surface gasping for air. If the pH is way off, do not rapidly restore the pH to normal. Instead, gradually add pH buffers until the proper pH is reached. To help prevent this problem, check the pH on a regular basis.

Oxygen Shortage

A shortage of oxygen can be diagnosed by rapid gill movement and fish hanging just below the water surface. Later, the fish may lose color and die. An oxygen shortage can be caused by several ways: insufficient aeration, a buildup of organic wastes, a high temperature, or through plant respiration. An oxygen deficiency can be solved by a partial water change, an increase in aeration, and removal of dead or dying fish and vegetation (Kabata, 1985)

Living Sources of Stress

Sources of stress to fish include living pathogens such as parasites, bacteria, fungus, and viruses (least common). These pathogens are most likely to affect fish when other conditions are weakening the fish, or when a new, infected fish is introduced. Other fish also serve as sources of stress when they attack other fish. “S” = symptoms; “A” = action.

COMMON PARASITES AND DISEASES AFFLICTING TROPICAL FRESHWATER SPECIES PARASITES

Velvet Disease (Oodinium)

S: Velvet disease is characterized by a cover of a fine gold to gray film on the fish's body. The affected fish may gasp for air and rub against rocks. The gills may be flared.

A: When Velvet disease has been confirmed, raise the water temperature of the tank. If possible, the temperature should be brought to 88-93°F (31-34°C). Turn off the lights and treat fish with copper sulfate, Trypaflavine, methylene blue, a malachite green-formalin combination, or Quinine Hydrochloride. An alternative is to immerse the infected fish in a salt bath.

White Spot Disease, Ich (Ichthyophthirius)

S: The body and fins are covered with small white spots. In heavy infestations, the skin may be covered with slimy gray patches. As the disease progresses, the fish becomes emaciated and less active, and scratches against objects.

A: The Ichthyophthirius parasite has three life cycle stages, and is only vulnerable to medication in the free-swimming stage. Raise the temperature to 86°F (30°C). This highly contagious disease is treatable with a malachite green-formalin combination, Trypaflavine, Quinine, an addition of salt, or one of many medications available in pet stores. Treat the tank for three weeks so all the Ich parasites have completed their cycles. Survivors appear to be immune after infection, but still carry disease. Non- immune tank mates are infected when they are weakened by stress.

Gill Flukes (Dactylogyrus, Cichlidogyrus, Tetraonchus)

S: Tiny worm-like flukes infest the gill membranes causing redness and slimy gills, panting at the surface, rapid breathing, and emaciation.

A: Flukes lay eggs that are resistant to medication. Treat the tank with Droncit (Praziquantel) at 2 ppm (2 Mg/L) or a malachite green-formalin combination until all eggs and flukes are gone. Flukes can be eliminated from fish (but not the tank), by short formalin, salt (3 %), or Ammonium Hydroxide baths.

Skin Flukes (Gyrodactylus)

S: Fish are infested with small worms causing colors to fade, reddish patches, and the fish to scratch against objects.

A: Skin flukes can be treated with Droncit (Praziquantel) at 2 ppm (2 Mg/L) and formalin baths.

Fish Lice (Argulus)

S: Fish rub against rocks and plants and have clamped fins. Areas may have red, inflamed spots. Small lice are visible to the naked eye.

A: The best treatment is careful, manual removal with a pair of tweezers and siphon the gravel daily to remove eggs. If there is a heavy infestation, raise the temperature to 86°F (30°C), and treat the aquarium with dylox, masoten, or trichlorfon (0,0-dimethyl- 2,2,2-trichloro-1-hydroxyethyl phosphonate). Be sure to remove all invertebrates from the aquarium as they may be harmfully affected by the treatment.

Costia

S: The body is covered by a gray layer of slime and the fins are frayed. The fish may swim erratically and rub against rocks. Heavy infestations result in reddish patches on the skin. A highly infectious disease that thrives in acidic water.

A: The temperature should be raised to 86°F (30°C) and the fish should be bathed in a short formalin, or a longer salt bath. Medications are available in pet stores.

Chilodonella

S: The body and gills lose color and become gray-blue in color. Fish swim erratically and rub against objects.

A: A malachite green-formalin combination works well as does a short (30 min) salt bath followed by a 12 hour Trypaflavine bath. This disease is most prevalent in overcrowded tanks and is most easily prevented by maintaining an appropriately stocked tank.

Hole-in-the-Head Disease, Discus Disease (Hexamita, Spironucleus)

S: Diseased fish lose weight and develop "pitting" in the head region.

A: The cause of this disease is usually attributed to a lack vitamins and minerals, especially vitamin D and calcium, the effects of the deficiency possibly amplified by the presence of Hexamita in the intestines. This intestinal flagellate is usually introduced with the feeding of Tubifex worms, and can survive in the gravel of unclean tanks. The best way to cure Hole-in-the-Head Disease is to complement the diet with vitamins. Treatment can also include dosages of antibiotics and metronidazole. Keeping the tank scrupulously clean also helps prevent this disease. Try medicated fish foods (Kabata, 1985)

FUNGAL DISEASES

Saprolegnia and Achlya

S: Cotton-like tufts of fungus appear on the body of the fish. As algae grows on the fungus, the fungus turns a brownish color.

A: Fungal infections are secondary infections that can only occur when the fish is already diseased or physically injured. When fungal infestations occur, raise the water temperature. Several different courses of action can be taken including a 30 minute bath in 1 ppm Sodium Permanganate (1 Mg/L); the addition of Malachite green; or the addition of a commercial fungal medication.

BACTERIAL INFECTIONS

Fish Tuberculosis (Mycobacterium)

S: Fish may lose color and appetite and become hollow-bellied. Fish become lifeless and often crippled—with a bent spine. Fish develop ulcers under the skin and may rupture causing open sores and "pop-eye."

A: This disease is highly infectious and deleterious. Bacteria can remain living in the gravel to infect other fish when they are weakened. Some success has been achieved by treating infected fish with antibiotics (Oxytetracycline and Kanamycin), although often it is best to kill the fish and put it out of its misery. When removing piscine tuberculosis victims, do so with care, as this disease can be transmitted to humans.

Dropsy (Aeromonas, Pseudomonas)

S: Fish infected with Dropsy are characterized by protruding scales, bulging eyes, pale gills, the body cavity is filled with fluid, and there may be red patches on the skin.

A: An infectious disease that spreads most readily among weakened fish. Infected fish should be removed and destroyed, or treated with antibiotics. Treatment is not usually successful.

Fin Rot (Pseudomonas and others)

S: The edges of the fins are discolored and frayed. As the disease progresses, fin damage becomes more evident as the fins disintegrate. Often fungal infections follow fin rot, contributing to fin damage.

A: Fin rot is most commonly caused by improper water conditions including too low a temperature and the buildup of toxic compounds. This infectious disease can be treated with a bath in Trypaflavine or the addition of commercially prepared medications. Treatment of Fin rot is difficult.

Mouth Fungus, Columnaris (Chondrococcus, Cytophaga)

S: Despite its misleading common name, mouth fungus is actually caused by bacteria. Patches of cotton-like patches develop around the mouth, but also on the head, fins, gills, and body. As the disease progresses, open sores develop.

A: Raise the water temperature. Several different courses of action can be taken including a 30 minute bath in 1 ppm potassium permanganate (1 Mg/L);

the addition of Malachite green; or the addition Nifurpirinol. Frequent partial water changes are important.

Neon Disease (Sporozoasis)

S: This incurable disease can affect characins, Cyprinds, and cichlids. The disease can manifest itself in several ways. Symptoms vary, and can include a loss of color, emaciation, and the loss of equilibrium causing fish to swim in an erratic, jerky manner. An infected specimen will wander from its school. The body may become a milky, opaque color.

A: Since this disease is not treatable, the best way to prevent its spread is to immediately remove the affected fish, and destroy it. Disinfect the tank after removing other fish.

Discus Flu, Discus Plague

S: The disease sets in rapidly after new fish are introduced. The first symptoms include small white patches on the body and the disintegration of the fins. Soon, the milky mucous membrane begins to slough off in large sections and the fish turn a dark color. If a number of fish are affected, they may huddle together in the corner of the tank.

A: This seasonal, flu-like disease, that has devastated entire hatcheries, is caused by seemingly unknown causes. Scientists have found a number of species of bacteria in diseased fish. The best way to combat the disease is to keep the pH low (4-5), make daily water changes (clean the tank well), and stop feeding the fish. The lights should be turned off and decorations removed. Try treating the water with a small amount of potassium permanganate to kill off some bacteria. Some suggest using an antihistamine throughout the course of the disease (Landsberg,1989)

ABOUT CITY WATER

City water and well water may contain chemicals which are toxic to freshwater aquarium fish. START RIGHT neutralizes most foreign chemicals found in water, such as bromine, chlorine, fluorine and iodine and will neutralize many harmful heavy metals. For this reason, it is wise to use START RIGHT with every partial or total water change.

Nitrites, which are found in all aquarium water, cause a fish blood condition known as methemoglobinemia. Methemoglobin (brown blood) cannot carry sufficient amounts of oxygen, and so nitrites slowly rise, depriving the fish of oxygen. New fish may die in a short time when added to nitrite laden water. START RIGHT contains special chloride ions which help guard fish against nitrite poisoning at up to twenty times normal levels.

Tropical Fish Problem Solver Chart

Please Note: We recommend you remove any activated carbon from your filter(s) when treating fish diseases.

A B C D E F G H I J K L M N O P Q R S T U V W X Y Z

Problem	Appearance	Cause	Proper Action	Notes
ACID WATER	Acidic water can irritate gills and shin. Fish may dart and gasp.	Acid from food waste or decayed food.	Use pH INCREASER. Check pH with test kit	To reduce acid buildup, change 10% of the water by siphoning from bottom of tank once a week. Add pH STABILIZER to maintain desired pH.
ADDING NEW WATER	Fish may act distressed.	Chlorine, chloramines, copper or ammonia in the water	Use START RIGHT if chlorine is present. Use ACE if ammonia or chloramines are present.	START RIGHT neutralizes chlorine, copper, heavy metals and harmful gases instantly. Guards against nitrite poisoning. Stimulates natural slime coat on fish.
ALGAE OUTDOORS (GREEN SLIME)	Green mat on bottom or the water is green.	Light too strong. Water too alkaline.	Use POND BLOCK. Add pH DECREASER	Shade pool if possible. Keep pH 7 or better pH 6.5 with pH DECREASER. Add pH STABILIZER to maintain desired pH.
ALKALINE WATER	White water line. Plants die, fish have frayed fins. Alkaline water can destroy gill and fin tissues. Ammonia becomes more toxic.	Seashells or limestone in tank or hard water supply.	Use pH DECREASER or NATURAL COLORS. Use a JUNGLE WATER SOFTENER.	A JUNGLE WATER SOFTENER will discourage green slime and keep white water line from forming on the glass. Add pH STABILIZER to maintain desired pH.
AMMONIA IN WATER	Fish die. Water may be cloudy.	Product of decaying wastes.	Use AMMONIA SORB, AMMONIA SORBER, or ACE.	To lower high ammonia levels use ACE AMMONIA CHLORAMINE ELIMINATOR
ANCHOR WORM	Threadlike worms hanging from fish. Base of anchor worm may be red.	A parasitic crustacean. (Lernaea)	Use PARASITE GUARD.	Use heavy aeration, as dead parasites may otherwise cause water to become cloudy.

Contd...

Problem	Appearance	Cause	Proper Action	Notes
ARGULUS	Quarter inch bubble like lumps slowly moving on skin and fins.	A parasitic crustacean. Sometimes Ergasilus also.	Use PARASITE GUARD.	Use heavy aeration, as dead parasites may otherwise cause water to become cloudy.

Problem	Appearance	Cause	Proper Action	Notes
BETTA PINCH FIN	Fish floats with fins closed.	Gill parasites.	Use PARASITE GUARD.	Use heavy aeration, as dead parasites may otherwise cause water to become cloudy.
BLACK SPOT	Small black specks on skin.	Small fish worms (Diplostomulum)	Use PARASITE GUARD.	Use heavy aeration, as dead parasites may otherwise cause water to become cloudy.
BLOODY FINS	Red streaks on fins, especially goldfish.	Gyrodactylus parasites (monogenetic trematodes*).	Use PARASITE GUARD.	Use heavy aeration, as dead parasites may otherwise cause water to become cloudy.

Problem	Appearance	Cause	Proper Action	Notes
CHLORAMINE IN CITY WATER	Fish dying within 24 hours due to blood poisoning. Gills become brown.	Water company adding chloramine as a substitute for chlorine.	Use ACE-AMMONIA CHLORAMINE ELIMINATOR	ACE is a safe and easy way to eliminate chloramine, ammonia and chlorine. Use with START RIGHT to fully condition the water.
CHLORINE IN WATER SUPPLY	Fish darts about. Gills pump furiously.	Chlorine added to water supply to kill bacteria.	Use START RIGHT	START RIGHT neutralizes chlorine, copper, heavy metals and harmful gases instantly. Guards against nitrite poisoning. Stimulates natural slime coat on fish.
CLAMPED FINS	Fins held close to body, especially in livebearing fish.	Gill or body parasites.	Use PARASITE GUARD.	Use heavy aeration, as dead parasites may otherwise cause water to become cloudy.

Contd...

Problem	Appearance	Cause	Proper Action	Notes
CLOUDY WATER INDOORS	Water not crystal clear in established tanks.	Wastes, small animal and plant life. Possibly overfeeding.	Use CLEAR WATER.	Good filtration improves clarity. Check for overfeeding or overlighting. Install JUNGLE MIRACLE UNDER GRAVEL FILTER.
CLOUDY WATER IN NEW TANK	Small particles suspended in the water causing a cloudy look.	Floating silts from gravel especially a newly setup tank.	Use MAGI CLEAR.	Use CLEAR WATER in old established tanks to clear odors due to overfeeding.
CLOUDY WATER UTDOORS	Water not crystal clear.	Wastes, small animal and plant life.	Use POND BLOCK.	Direct sunlight should be avoided if possible. Shade pools with plantings or awnings.
COLUMNARIS	White cotton sprouting from fins or body.	A bacteria, Flexobacter (Chondrococcus).	Use FUNGUS GUARD or FUNGUS ELIMINATOR.	Can enter other aquariums by splashing water or using nets from infected aquarium. Clean nets with NET SOAK.
COSTIASIS	Grey film on all or part of body.	External protozoan parasite (Costia necatrix)	Use FUNGUS ELIMINATOR or BINOX. Use GOLD FIX on goldfish.	Can enter other aquariums by splashing water or using nets from infected aquarium. Clean nets with NET SOAK.
DACTY-LOGYRUS	Gills hang open at all times.	Small gill parasites (monogenetic trematodes*)	Use FUNGUS ELIMINATOR or PARASITE GUARD. Use GOLD FIX on goldfish.	Can enter other aquariums by splashing water or using nets from infected aquarium. Clean nets with NET SOAK.
DIRTY AQUARIUM SAND	Fish stir up waste from bottom and tank looks untidy.	Overcrowded tank or over feeding.	Use CLEAR WATER. Clean aquarium when convenient.	Use SERI CLEAN to clean tanks for less work and a better job. Use LIME OFF on lime deposits.

Contd...

Problem	Appearance	Cause	Proper Action	Notes
closely HEAD WORMS	wildly and may smash into glass.	protozoa in sinus canals of head.	HEAD GUARD.	during dip and remove when distressed, usually after a few minutes.
DROPSY	Fish bloated with scales standing out. Fish resembles a pine cone.	Kidney damage.	Use FUNGUS ELIMINATOR.	Kidney damage is usually the result of disease or too strong drug use.
Problem	**Appearance**	**Cause**	**Proper Action**	**Notes**
FIN FUNGUS	Fins ragged with jelly at end of ragged part.	Water change. This is not a disease.	Use FUNGUS GUARD.	Fin fungus is very rare when START RIGHT has been added in setting up the aquarium or during water change.
FIN ROT	Fins eaten away with redness at base of bad area.	Saprolegnia fungus with Pseudomonas or Aeromonas bacteria.	Use FUNGUS ELIMINATOR or BINOX.	Can enter other aquariums by splashing water or using nets from infected aquarium. Clean nets with NET SOAK.
FLAVO BACTERIA	Fish swim aimlessly. May have spasms.	Internal bacteria	Use INTERNAL PARASITE GUARD or BINOX with PEPSO FOOD.	Can enter other aquariums by splashing water or using nets from infected aquarium. Clean nets with NET SOAK.
FLESH WORMS	Small lumps under skin or on fins.	Parasitic worm	Use BINOX or FUNGUS ELIMINATOR. Use GOLD FIX on goldfish.	Will not spread indoors.
FUNGUS (white)	A white cottony growth or patch on any part of the fish.	A bacteria, Flexobacter (Chondrococcus) Columnaris	Use FUNGUS ELIMINATOR or GUARD. Use GOLD FIX on goldfish.	Can enter other aquariums by splashing water or using nets from infected aquarium. Clean nets with NET SOAK.
FUNGUS (grey)	A light grey cottony growth or patch on any part of the fish.	A parasitic fungus. Saprolegnia	Use FUNGUS GUARD with ICK GUARD.	Can enter other aquariums by splashing water or using nets from infected aquarium. Clean nets with NET SOAK.

Contd...

Problem	Appearance	Cause	Proper Action	Notes
FURUN CULOSIS	Open sores on Gouramis or other anabantids.	Aeromonas salmonicida bacteria.	Use FUNGUS ELIMINATOR.	Seldom spreads to other anabantids* and then only on direct contact.
GLASS DIRTY ON OUTSIDE.	Fingerprints or drip marks.	Finger grease.	Use SERI CLEAN.	Use full strength and wipe dry with paper towel for lasting resistance to stains or marks of any type.
GILL FLUKES	Gills hang partly open and may pump fast. Gills may appear red.	Small gill parasites (Gyrodactylus) monogenetic trematodes*.	Use PARASITE GUARD.	Use heavy aeration, as dead parasites may otherwise cause water to become cloudy.
GOURAMI DISEASE	Fish rocks back and forth while staying in place.	Dactylogyrus parasites (monogenetic trematodes*)	Use PARASITE GUARD.	Use heavy aeration, as dead parasites may otherwise cause water to become cloudy.
GREY SKIN	Skin looks grey in patches.	Costia necatrix parasites.	Use BINOX or FUNGUS ELIMINATOR. Use GOLD FIX on goldfish.	Can enter other aquariums by splashing water or using nets from infected aquarium. Clean nets with NET SOAK.
GYROD-ACTYLUS	Fish scrape on objects in tank.	Gyrodactylus parasites (monogenetic trematodes*)	Use PARASITE GUARD.	Use heavy aeration, as dead parasites may otherwise cause water to become cloudy.
HEMOR-RHAGIC SEPTICEMIA	Red streaks on fins usually near body with no sign of skin damage.	A bacteria, Aeromonas (liquifaciens) hydrophylia	Use FUNGUS ELIMINATOR. Use GOLD FIX on goldfish.	Common on goldfish and livebearing tropicals. Scales may sometimes protrude.
HYDRA	Tiny tree-like animals attached to objects in the tank.	Carnivorous microanimal.	Use SHIMMY BLOCKS or MOLLY BRIGHT.	Harmless to all except newly hatched fry, which may be eaten by hydra.

Contd...

Problem	Appearance	Cause	Proper Action	Notes
ICH	White sugar-like crystals on fins or body of fish.	Ichthyophthirius multifiliis, a parasitic protozoan.	Use ICK GUARD. Use ICK GUARD II FOR SCALE-LESS FISH on weak, small or scaleless fish.	Can enter other aquariums by splashing water or using nets from infected aquarium. Clean nets with NET SOAK.
INTERNAL PARASITES	Fish are thin and act listless.	Tiny internal parasites.	Use INTERNAL PARASITE GUARD or PEPSO FOOD.	Seldom spreads indoors.
INTRO-DUCING NEW FISH	Fish may go to top or bottom or dash about.	Change in water from which fish are acclimated.	Use START RIGHT.	START RIGHT neutralizes chlorine, copper, heavy metals and harmful gases instantly. Guards against nitrite poisoning. Stimulates natural slime coat on fish.
IRON IN WATER SUPPLY	Stains on glass or ornaments.	Dissolved iron in water supply.	Use CLEAR WATER. Use WATER SOFTENER.	CLEAR WATER neutralizes iron and WATER SOFTENER removes iron up to 1 1/2 parts per million.
ITCHING (also called FLASHING)	Fish scratches on objects in tank.	Skin irritation caused by tiny skin parasites.	Use PARASITE GUARD, FUNGUS ELIMINATOR or ICK GUARD.	Can enter other aquariums by splashing water or using nets from infected aquarium. Clean nets with NET SOAK.

Problem	Appearance	Cause	Proper Action	Notes
LIME DEPOSITS ON GLASS	Insoluble white deposit or water line.	Caused by lime, a mineral found in water supplies or from slat buildup.	Use LIME OFF.	Use ALGAE REMOVER or AQUARIUM CLEANING KIT for glass aquariums. Use ALGAE REMOVER from acrylic or plastic tanks.
LIVE BEARER DISEASE or MOLLY DISEASE	Fish stay in one place and wiggle, wag or shimmy.	Water lacks electrolytes needed for normal met abolism.	Use START RIGHT with either MOLLY BRIGHT or SHIMMY BLOCKS.	Not a disease at all but a reaction to water lacking needed electrolytes.

Contd...

Problem	Appearance	Cause	Proper Action	Notes
MOUTH FUNGUS	White cotton sprouting from mouth sides.	A bacteria, Flexobacter (Chondrococcus).	Use FUNGUS GUARD or FUNGUS ELIMINATOR.	Can enter other aquariums by splashing water or using nets from infected aquarium. Clean nets with NET SOAK.
MOUTH ROT	Flesh eaten away around mouth.	Saprolegnia fungus with Pseudomonas or Aeromonas bacteria.	Use FUNGUS ELIMINATOR or BINOX.	Can enter other aquariums by splashing water or using nets from infected aquarium. Clean nets with NET SOAK.

Problem	Appearance	Cause	Proper Action	Notes
NEON AND CARDINAL DISEASE	Color fades from red line, line may turn yellow.	A parasitic protozoan, Pilistophora hyphesobryconis.	Use CLEAR WATER or INTERNAL PARASITE GUARD with NATURAL COLORS.	Can enter other aquariums by splashing water or using nets from infected aquarium. Clean nets with NET SOAK.
NETS NEED CLEANING	Nets used between multiple aquariums may need to be cleaned between use.	Problems can spread from tank to tank.	Use NET SOAK to take the guesswork out of cleaning nets.	NET SOAK is also effective to clean ALGAE REMOVER SPONGES.
NEW AQUARIUMS	Dusty or oily.	Manufacturing residues.	Use SERI-CLEAN or AQUARIUM CLEANING KIT (SERI CLEAN with ALGAE REMOVER). Use ALGAE REMOVER for acrylic or plastic tanks.	SERI CLEAN washes much cleaner and easier than plain water. Install a new UNDER GRAVEL FILTER.
NEW FISH	Act uncomfortable or nervous.	Different water than the fish is accustomed to living in.	Use START RIGHT.	Electrolytes vary in water supplies. Add START RIGHT to lessen shock and add beneficial electrolytes.
NEW AQUARIUMS	Dusty or oily.	Manufacturing residues.	Use SERI-CLEAN or JUNGLE AQUARIUM CLEANING KIT (SERI CLEAN with ALGAE REMOVER). Use ALGAE REMOVER for acrylic or plastic tanks.	SERI CLEAN washes much cleaner and easier than plain water. Install a new UNDER GRAVEL FILTER.

Contd...

NEW FISH	Act uncomfortable or nervous.	Different water than the fish is accustomed to living in.	Use START RIGHT	Electrolytes vary in water supplies. Add START RIGHT to lessen shock and add beneficial electrolytes.
NEW GRAVEL	Dusty or powdery looking.	Residues from manufacturing processes or shipping abrasion.	Use GRAVEL WASH.	All gravel should be washed thoroughly before placing in an aquarium.
NEW WATER	Fish darts about. Gills pump furiously.	Dissolved metals or gases in the water.	Use START RIGHT.	Chlorine, copper and other heavy metals may often be found in city water. START RIGHT removes or neutralizes these chemicals at once.

Problem	**Appearance**	**Cause**	**Proper Action**	**Notes**
ODOR FROM WATER	Water may be cloudy.	Decaying food or plants.	Use CLEAR WATER. Use AMMONIA SORB.	Reduce amount fed per feeding. Partial water change should be done as soon as convenient. Install UNDER GRAVEL FILTER. Ammonia may still be present.
OODINIUM (also called VELVET or RUST)	Body skin of fish appears dusty in yellowish or grey patches.	A parasitic algae, Oodinium limneticum.	Use MOLLY BRIGHT or SHIMMY BLOCKS with VELVET GUARD.	Spreads easily. Keep aquarium as dark as possible, as Oodinium needs light to spread. Clean nets with NET SOAK.

Problem	**Appearance**	**Cause**	**Proper Action**	**Notes**
pH BOUNCES BACK UP AFTER BEING LOWERED	White water lines. Plants die, fish have frayed fins.	Hard water . supply Seashells or limestone in the tank.	Use pH STABILIZER to maintain desired pH.	Natural buffers in hard water interfere with the stability of pH once lowered.
PLANTS DIE	Leaves shed, stems rot, roots die.	Water unsuited for plants.	Use PLANT SAVER.	Plants as well as fish require conditions similar to their natural environment. Needed electrolytes are supplied by PLANT SAVER.

Contd...

Problem	Appearance	Cause	Proper Action	Notes
POP EYES	Eyes stand out from socket.	Infection from fighting.	Use FUNGUS GUARD or FUNGUS ELIMINATOR. Reduce light.	Fish often fight when ready to mate. Remove the pairs seeming to fight and move them to a special spawning aquarium.
RASBORA DISEASE	Rasboras die off in the aquarium.	Water unsuitable for fish.	Use NATURAL COLORS.	Rasboras and many tetras need the humic acids supplied by NATURAL COLORS. This is not a disease but a water deficiency.
ROT	Fish decays at the mouth or tail.	Saprolegnia fungus with Pseudomonas or Aeromonas bacteria.	Use FUNGUS ELIMINATOR or BINOX.	Extremely easy to spread. Be careful with nets and splashing to avoid spreading this to other aquariums. Clean nets with NET SOAK.
SAPRO-LEGNIA	Frayed looking fins and moldy looking slime on body.	A parasitic fungus.	Use FUNGUS GUARD with ICK GUARD.	Spreads easily. Keep tank as dark as possible during treatment as Saprolegnia needs light. Clean nets with NET SOAK.
SHAKES AND SHIMMIES	Fish wags and shakes while standing still.	Water lacks electrolytes.	Use MOLLY BRIGHT or SHIMMY BLOCKS.	This is not a disease but a warning that something is lacking in the water used. MOLLY BRIGHT and SHIMMY BLOCKS supply the needed electrolytes.
SHIPPING FISH LOSSES	Fish act stunned or gasp at top of water.	Ammonia and carbon dioxide suffocating fish in container.	Use START RIGHT and ACE-AMMONIA CHLORINE ELIMINATOR.	Ammonia from fish waste and carbon dioxide from fish breathing are combined and neutralized.
SMALL FISH DEATHS	Small fish, especially neons, angels and guppies die without visible symptoms.	A true fungus called Saprolegnia invades the gills causing suffocation.	Use SMALL FISH SAVER.	SMALL FISH SAVER will also help protect fish eggs from fungus.

Contd...

SMALL TETRA LOSSES	Fish won't eat. Heavy loss of thin specimens.	Water unsuitable, as many of these fish are caught wild.	Use CLEAR WATER and NATURAL COLORS.	Fish raised in the wild need the humic acids supplied by NATURAL COLORS.
SORE EYES	White film on eyes.	Eye injury. From handling or breeder fighting.	Use FUNGUS ELIMINATOR or BINOX.	Check for breeding pairs if there are older fish. Separate these and place in spawning aquarium.
SOUR WATER	Water cloudy and may smell bad.	Rotten food or imbalance caused by strong drugs.	Use CLEAR WATER. Install JUNGLE MIRACLE UNDER GRAVEL FILTER.	Reduce feeding amounts if overfeeding. If from strong drugs, the smell comes from losing the useful bacteria that balance the water.
STARTING NEW AQUARIUMS	Fish act very uncomfortable.	Chlorine, chloramine, copper or manufacturing oils in the water.	Use START RIGHT. Use ACE along with START RIGHT if chloramine is present.	START RIGHT neutralizes chlorine, copper, heavy metals and harmful gases instantly. Guards against nitrite poisoning. Stimulates natural slime coat on fish.
STARTING NEW POOLS	Fish show blood on fins or die.	Strong alkali from new cement.	Use pH DECREASER with START RIGHT.	Two ounces of pH DECREASER will neutralize the cement alkali of each 200 gallon capacity. Test with test kit. Use pH STABILIZER to maintain desired pH.
SWIM BLADDER DISEASE	Fish swimming with head down (not level).	Bladder infection.	Use FUNGUS ELIMINATOR.	Does not spread to other fish.

Problem	Appearance	Cause	Proper Action	Notes
TAIL ROT	Tail eaten away with red area next to body.	Saprolegnia fungus with Pseudomonas or Aeromonas bacteria.	Use FUNGUS ELIMINATOR or BINOX.	Spreads very easily. Use caution to avoid splashing or using nets from infected aquarium. Clean nets with NET SOAK.

Contd...

THIN FISH	Fish thin and weak	Internal or external parasites. See GYROD-ACTYLUS.	Feed PEPSO FOOD and use INTERNAL PARASITE GUARD.	Use care to avoid spreading to other aquariums. Clean nets with NET SOAK.

Problem	**Appearance**	**Cause**	**Proper Action**	**Notes**
UNCLEAN GRAVEL	Green or brown film coating gravel.	Algae buildup, fish waste or uneaten food.	Use JUNGLE GRAVEL WASH.	Dirty gravel may lead to undesirable conditions.

Problem	**Appearance**	**Cause**	**Proper Action**	**Notes**
VELVET	Body skin of fish appears dusty in yellowish or grey patches.	A parasitic algae, Oodinium limneticum.	Use MOLLY BRIGHT or SHIMMY BLOCKS with VELVET GUARD.	Spreads easily. Keep aquarium as dark as possible, as Oodinium needs light to spread. Clean nets with NET SOAK.

Problem	**Appearance**	**Cause**	**Proper Action**	**Notes**
WHITE SKIN	Grey or white film on body.	External protozoan parasite (Costia necatrix)	Use FUNGUS ELIMINATOR or BINOX. Use GOLD FIX on goldfish.	Spreads very easily. Use caution to avoid splashing or using nets from infected aquarium. Clean nets with NET SOAK.
WHITE SPOT (ICH)	White sugar-like crystals on fins or body of fish.	Ichthyophthirius multifiliis, a parasitic protozoan.	Use ICK GUARD. Use ICK GUARD II FOR SCALE-LESS FISH on weak, small or scaleless fish.	Can enter other aquariums by splashing water or using nets from infected aquarium. Clean nets with NET SOAK.

Source: Roberts, 2001, Sarig, 1971, Sineszko, et al. 1970 and Srivastava, 1975.

4

Parasitism Versus Water Quality Parameters

INTRODUCTION

It can be added that physiological stress as well as the physical offences contribute as primary factor for disease occurrence and fish mortality. Many parasitic diseases due to fluctuation in water quality leads to deplition of stock, and some times total stock will go eg. yellow grub disease.

EFFECT OF TEMPERATURE ON PARASITES

Heterophyid Trematode

Cyprinodon variegatus, an estuarine fish, is heavily parasitized in the bulbus arteriosus by the heterophyid trematode *Ascocotyle pachycystis.* Parasitized fish ventricles were hypertrophied due to mechanical blockage of normal blood flow by metacercariae in the bulbus Swimming performance generally was reduced by the presence of parasites, which had their most significant negative effect at cold temperatures and low dissolved oxygen level (Felicia C. Coleman , 1993)

Myxosporean Parasite

Infection by *Ceratomyxa shasta* was occur only when water temperatures exceeded 10°C, thus accounting for the seasonal occurrence of the disease.Fish can become infected in water at temperatures as low as 4 to 6°C,although fish are infected at these lower temperatures, the progress of the disease is temperature dependent(Ratliff, 1983 and Ching and Munday,1984a).

Oviposition Rate In *Protopolystoma Xenopodis*

In reproductive capacity with parasite age, oviposition rate was severely depressed at lower temperatures. Egg production was monitored in a sample of seven hosts at 2°C intervals between 20°C and 8°C. Overall mean egg production rate showed a consistent decline from 11.9 eggs/worm/day (e/w/d) at 20°C (days 1–10) to 0.9 e/w/d at 8°C (J. A. Jackson and R. C. Tinsley 1997)

Sarcomastigophora

Effects of temperature on amyloodinium (sarcomastigophora, marine velvet disease): Optimum temparature for tomont division & sporulation ranges from 23-27°C. Completion of tomont division limited to 16-30°C. Infections do not occur less than 17°C. Effects of temperature on piscinoodium (sarcomastigophora). Life cycle completed in 10-14 days under optimal conditions. The optimum temperature for *P.pilluare* is 23-25°C & sporulation requiring 50-70 hrs for an average sized tomont. At 15-17°C sporulation require 11 days.

Ciliophora

Effects of temperature on *I.multiphilus* (ciliophora): At 25-28 °C parasite activity increases&life cycle completed in short time.The number of tomites in each cyst varies with temperature of the water (Woo, 2006)

Copepodes

Eragasilus (copepoda):egg sac formation stimulated in spring and early summer,but it is late in late summer &autumn. Lernea (copepoda) optimum temperature for development is 26-28°C. It is more prevalent in still/slow flowing waters than in fast flowing streams (Woo, 2006).

Parasitic Community

The prevalences of the epizoic ciliates *Trichodina* sp. and *Apiosoma* sp., and monogeneans *Dactylogyrus suecicus*, *Dactylogyrus nanus*, and *Gyrodactylus* sp. peak in April at a water temperature of 8°C. Prevalence of metacercariae of the digeneans *Diplostomum. spathaceum* and *Tylodelphys clavata* remained constant with values around 60% and 10%, respectively. Larvae of the nematode *Philometra obturans* and the cestode *Ligula intestinalis* and females of the copepod *Neoergasilus japonicus* were found sporadically. (Klaus Knopf *et al.*, 2007)

EFFECTS OF SALINITY ON PARASITES:

Sarcomastigophora

Effects of salinity on amyloodinium (sarcomastigophora, marine velvet disease).Effective tomont division and sporulation occurs upto 50ppt.Minimum effective salinity varies from 12-20ppt depending upon isolates.

Myxosporean Parasite

Only limited information is available about the effects of salt water on the progress of a *C. shasta* infection.Normally infections occurs below 15ppt.That infections were prevented at salt concentrations greater than 15ppt (Johnson, 1975).

Effects Of Salinity & Temperature On Copepodes

Parasite abundance increased inversely to water temperature and salinity with highest incidences occurring in fish from lower saline and colder waters. Body site preference, seasonality, abundance, and incidence of the parasite *Lernaeenicus radiatus* in relation to water temperature and salinity variations. Fins were preferred attachment sites in Atlantic croakers regardless of fish size (James *et al.* 1979)

Effects Of Salinity And Temperature On Fish Parasites

In brackish water the variety of marine and freshwater parasite species is considerably reduced. Most of the parasite species are more tolerant than their hosts. First stage larvae of *Contracaecum aduncum* develop in 0-32‰ salinity.*Cryptocotyle lingua* proved to be infective at salinities down to 4‰. Parasites in the fish intestines appear to be unaffected by changing water salinities, as the osmolarity in the intestines stays nearly constant. Marine ectoparasites (*Acanthochondria depressa, Lepeophtheirus pectoralis*) survive about three times longer than freshwater species (*Piscicola geometra, Argulus foliaceus*) when salinity is 16% (Möller, 2005)

Effect Of Decreased Water Oxygen Content On Common Carp Fry With *Dactylogyrus Vastator*

Rapid infection of small fish by large numbers of larvae very often causes mortality in common carp fry (Paperna, 1964 and Bauer *et al.* 1981). Fish can tolerate even infections of very high intensity; however, their survival chances are negatively affected by adverse environmental factors and poor feeding conditions (Schaeperclaus,1954).

EFFECT OF LOW pH ON *TRYCHOPHRYA INTERMEDIA*

Trychophrya intermedia on the gills of rainbow trout acclimating to low ambient pH. Rainbow trout under low pH conditions acquire heavy infections with

Trychophrya intermedia, not secondary to gill damage.The parasite may have a primary effect on gill function in fish under acid condition (Balm, *et al.*, 1995)

EFFECT OF pH and ELECTRIC CONDUCTIVITY ON I. *MULTIFILIS*

A significant negative correlation exist between *I. multifilis* and pH. Increased electric conductivity and pH provoked reduction of *I. multifiliis* infestation. Salt application in the water to increase the electric conductivity may also prove to be a feasible method of *I. multifiliis* control (GarciaI *et al.*, 2009).

Association Between Trichodiniasis In Eel (Anguilla Anguilla) And Water Quality In Recirculation Systems

Trichodiniasis had relatively high contents of organic dry matter (13–65 mg l”1), low concentrations of nitrate (150–500 mg NO3 l”1) and high pH (6.2–7.4) in the process. Infection pressure from trichodiniasis in farms with a relatively high load of organic dry matter (>15–20 mg l”1) may be reduced by reducing the content of organic dry matter in the process water (Hans and Madsen, *et al.* 2000).

Water Quality As A Determinant Of The Composition Of Fish Parasite Communities

The degree of interactivity among parasites, parameters of species richness, diversity suggest that the structure of parasite communities are affected by the water contamination level. Parasites are indicative of many biological aspects of their hosts, including diet, migration, recruitment and phylogeny they may also be good direct indicators of environmental quality status. Parasites can thus be considered complementary to chemical analysis or traditional biological surveys (bacteria counts or invertebrate assessments) as indicators of dysfunction at the ecosystem level. Eg: ectoparasite densities increased proportionally with the eutrophication level while endoparasites were restricted to unpolluted lakes (Galli, *et al.*,2001).

WHY PARASITES ARE IMPORTANT?

Entire parasite community as a biological indicator of the fish environmental conditions. Two bays are submitted to urban wastewater inputs alone or combined with additional industrial inputs. The encysted metacercariae found in the pericardic cavity were found to be significant indicators of the specific anthropogenically impacted environmental conditions prevailing in the inner parts of the two bays. (Sasala Pierre, *et al.*, 2007).

Parasites Of Fish As Indicators Of Environmental Stress

Environmental parameters which may affect parasite abundance, richness, prevalence, and diversity can be divided into three categories. 1) the physical and chemical characteristics of the water and sediment (including contaminants) external to the fish. 2) the internal environment defined by the physical condition (physiological) of individual fish. 3) the presence and relative abundance of benthic macro invertebrates. The biotic response of parasites to environmental stressors is also reflected in the health of fish. Parasite assemblages of silver perch *Bairdiella chrysura* respond to both natural and anthropogenic stressors. particular environmental stressors and specific parasites that respond include: Temperature and monogeneans; contaminants and nematodes; low dissolved oxygen and protists; and salinity, together with a mixture of metal and organic contaminants and crustacea. Parasites of fish are useful biomarkers and appear to be more sensitive to environmental stressors than fish themselves. Parasite responses to selected environmental stressors may be used to discriminate polluted and unpolluted sites. The use of parasites of fish as biomarkers has relevant application to fisheries management and coastal monitoring programs. (Landsberg, *et al.* 2006).

Parasites As Indicators Of Pollution

To test the usefulness of parasites to detect pollution, evaluated the parasite loads of common estuarine fish at polluted and unpolluted sites Tested for differences in parasites that have direct life cycles(one host, like a fish) and parasites that have indirect life cycles (multiple required hosts, such as first a snail and then a fish). The monogeneans studied have direct life cycles then parasite numbers should show an increase. The parasites that require multiple hosts (digeneans) would show a uniform and consistent decrease in abundance. The reduction in the numbers of these parasites would be caused mainly by pollution induced high mortality rates of intermediate hosts of these parasites.

Q. Parasite loads of fishes in polluted areas would increase or decrease Why?

If pollutants adversely affect parasites, then a decrease in their abundance should be seen. Conversely, if the hosts (fish) of these parasites are stressed by pollutants in a manner that reduces their ability to resist parasitic infections. To manipulate the physical environment at farm such as temperature, salinity, pH, and dissolved oxygen in such a way to minimize effective transmission of the parasite and thus minimize their impact (Font, *et al.*, 2007)

5

Machanism of Parasitic Attachment in Fish

INTRODUCTION

An association of two organisms, one living at the expense of other for food and shelter is called parasitism. Parasite factors associated with recognition and selection of the host and the mechanisms in the host responsible for acceptance or rejection of the invading organism were evaluated. Sensory structures in parasites are able to detect differences between different fish species and this ability to discern between fishes may be based on both chemical and mechanical stimuli on the host surface. Complex glycoproteins, proteins, carbohydrates and simple molecules attract parasites or modify their behaviour. (Kurt Buchmann and T. Lindenstrom, 2002)

- PROTOZOANS
- CILIATES: The scopular surface secretes a substance for attachment of the ciliate to the surface of its host. These cilia are firmly wedged among the microvilli of the surface of the host's body, assuring firm attachment of the ciliate(JiYí Lom and John O. Corliss, 1968) .Some examples are as follows;

Ichthyophthirius multifiliis : Invasion of tomites (theronts) (30–45 ¼m long) into the host integument is facilitated by the excretion of a sticky substance from subpellicular crystalline organelles named mucocysts. Active penetration causes focal necrosis of the epithelial cells. It has been suggested that hyaluronidase and other enzymes may be produced by the penetrating parasite (Ewing *et al.*, 1985). In the absence of a suitable host, tomites will lose their infective potential within 24 hours at 24–28°C (Ewing and Kocan, 1992).

Trichodina: Body shaped like hockey puck,also spiral cilia around cytostome, makes them fly through the water like a flying saucer

Ambiphrya ameiuri ; It has cylindrical shape, row of oral cilia, and middle bank of cilia

Epistylis: *Epistylis* is a stalked ciliate. Believed to secrete proteolytic ("protein eating") enzymes that create a wound, suitable for bacterial invasion.

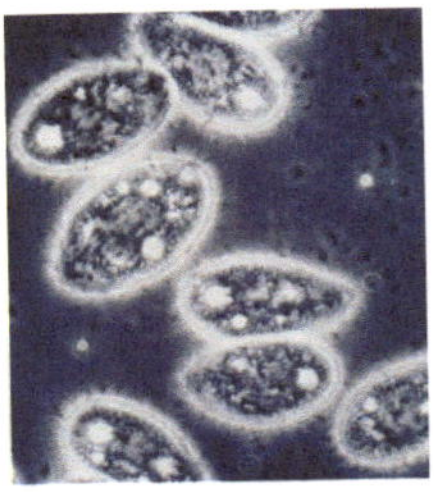

Ichthyophthirius multifiliis

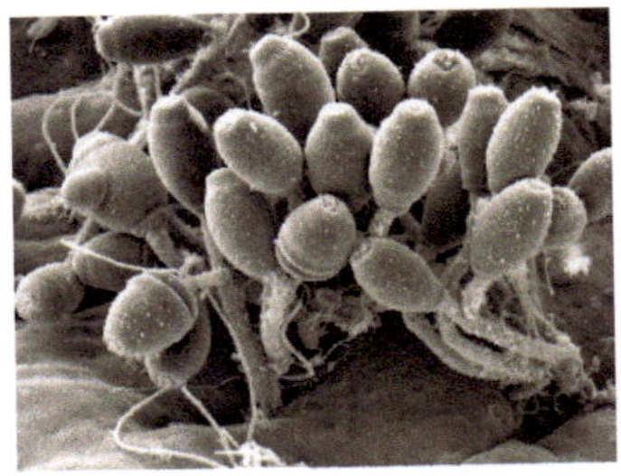

Epistylis

▪ Flagellates

Cryptobia: They are non-pathogenic and ectocommensal. Free, spindle shaped, 10–30 × 35 ¼m in size (if *C. branchialis*), or pyriform when attached to the integument, with two flagellae, one wholly or partly adjunct to the body

A prominent ridge on the flagellar membrane contacts the host cell membrane, forming a zone of gap junction and leaving a space of about 10 nm between the two membranes. The flagella do not penetrate into the epithelial cells (Lom, 2006).

▪ Dinoflagellates

Oodinium, Amylodinium: The parasitic stage (trophont) is spherical or pyriform.They are attached to and feeds from the host epithelial cell by means of rhizoids, which penetrate the host cell. The consumed cell gradually degenerates and collapses. Damage to infected cells leads to focal erosion of the epithelium. Prolonged infection exhausts a generation of mucus cells and leads to accelerated desquamation.

Proliferation of the epithelium causes obliteration of the gill lamellae, while the inner strata of the epithelium become spongious and in some cases undergo complete lysis (Paperna, 1980).

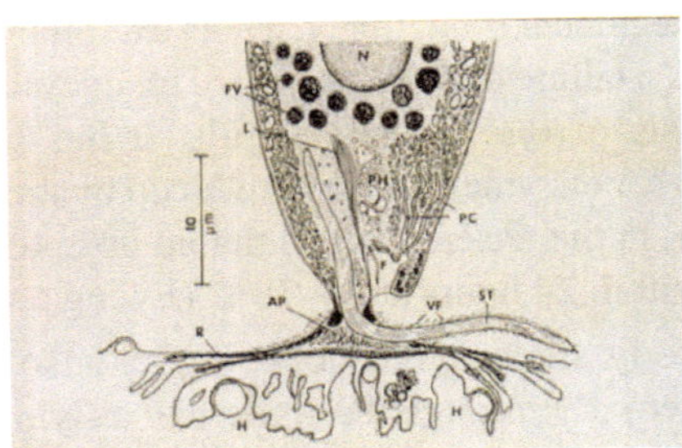

▪ Myxozoans

Myxobolus cerebralis: Histozoic infections are whitish cysts with a milky substance containing microscopic (5-5 ¼m longer axis) spores. The so called cyst is a parasite-origin plasmodium which forms a specialised membranous junction of pinocytotic vesicles (canals) with the surrounding host cells (Current and Janovy, 1976).

Spores, hard-walled, with one to six polar capsules are differentiated by sporogenesis within the plasmodium. Coelozoic myxosporea of the urinary and bile cavities have a small plasmodium and produce few spores, often only two. Premature plasmodia are attached by pseudopodia to the epithelial lining of the bladder. (Paperna *et al.*, 1987). It can cause "blacktail"(trouts) by controlling production of chromatophores in spinal column, and skeletal deformities.(VanWyk,1968).

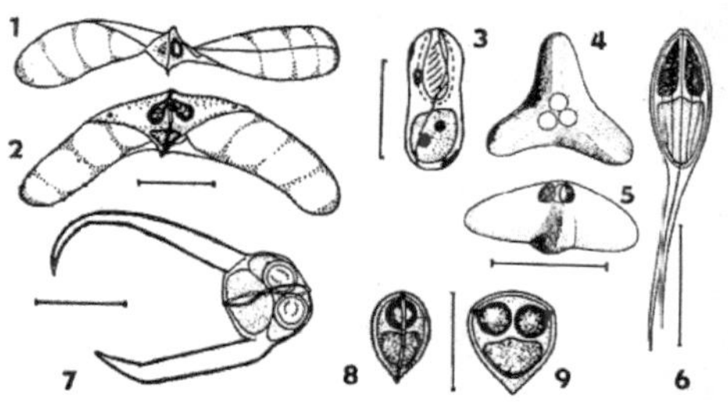

▪ Microsporidian

These unicellular eukaryotes exhibit a unique invasion mechanism. The spore, which is the infectious stage, contains an extrusion apparatus constituted by a polar tube inserted at its anterior end in an anchorage disk. Under the effect of certain stimuli, which can be linked in vitro to a variation in the pH, osmolarity, or the presence of cations or anions, the polar tube is extruded by the microsporidian spore and traverses the plasma membrane of a cellular host. The sporoplasm, expelled via this tube, is thereby inoculated into the receptor cell. These invasive apparatus, specific to the microsporidians and unique in the world of living organisms. Microsporidian spore invasion tubes form by the eversion of polar filament protein (PFP) and presumptive extrusion apparatus (EAP) membrane (Cali, 1991).

▪ Coccidia

These are intracellular parasites, mainly found in intestine. Coccidians are identified by the morphometry of their oocysts and sporocysts, the site of endogenous development and their position in the host cell (Dykova & Lom, 1981).

The sporocyst's hard wall is either bivalved and cleaves by a longitudinal suture (in the genus *Goussia*, Paperna & Cross, 1985), or opens at the sporocyst's apex, at one pole, through a round pore (genus *Epieimeria* of eels, and some *Eimeria* s.l.) or a short apical suture. The sporocyst wall of the latter type may also form tubercules or projections and is further enclosed in a veil (genus *Calyptospora* , Overstreet *et al.*, 1984).

▪ MONOGENETIC TREMATODES (FLATWORMS)

Dactylogyrus/ Gyrodactylus:

Dactylogyroidea are 0.3–2 mm long and usually have one or two anterior-dorsal pairs of eyes and a posterior-ventral attachment organ (the opisthaptor). This disk-like organ contains centrally positioned sclerotinoid anchors, connected to support bars and marginally located hooklets. Most dactylogyrids are gill parasites only a few are skin parasites, most notably Gyrodactylidae. The genera *Dactylogyrus*, parasitic in cyprinids, have one pair of anchors.

Attachment of the haptor was achieved mainly by the blade of the hamuli, but no evidence was found indicating the participation of marginal hooklets in the attachment. The hamuli penetrate into the interlamellar epithelium of the gill tissue. Some evidences were found to indicate that *they* may also utilize suction force during haptoral attachment. (Safaa Z. Arafa ,Mohammed M. El-Naggar and Samir A. El-Abbassy 2009)

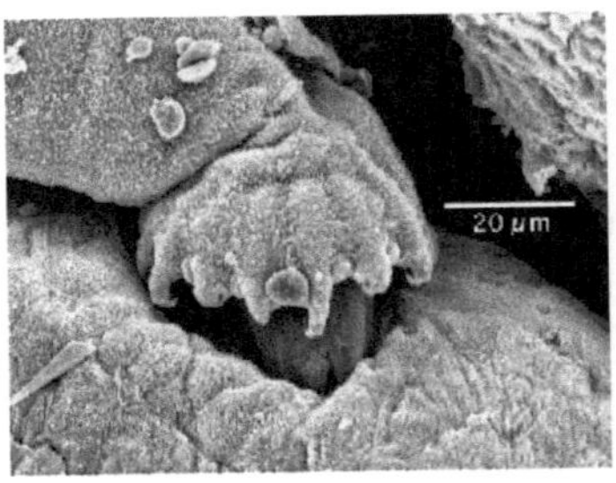

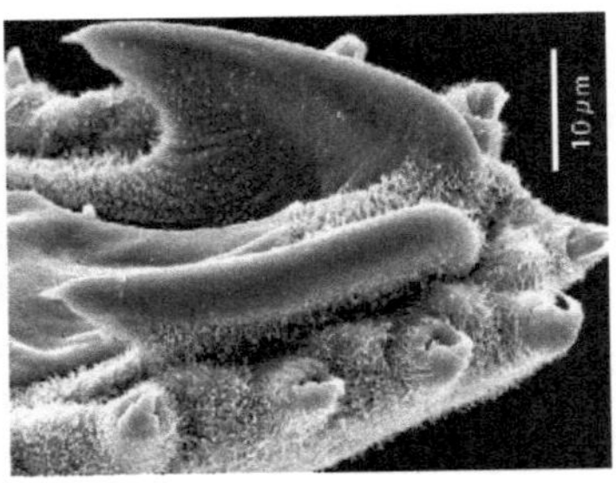

▪ Digenetic Trematodes

Adult-stage digeneans usually have a dorso-ventrally flattened, oval body with a smooth, spiny or corrugated surface, a sucker around the antero-ventral mouth, and an additional ventral sucker or acetabulum. Both suckers are used for attachment and locomotion (Schell, 1970).

Sanguinicola (Blood fluke) : In Sanguinicolidae and Plagiorchidae (Hoffman, 1960 and Smith, 1972), the cercariae are formed in the daughter sporocyst stage. Cercariae already have the elements of mature digenean organisation, but with primordial genital organs. Cercariae may also have locomotive devices; a tail, in some forked, fins (the forked-tailed cercariae of sanguinicoliids also have a characteristic dorso-median fin fold), and a pair of eyes. The latter are lost when cercariae transform into metacercariae.

Of all piscine trematodes, only the blood flukes (Sanguinicolidae) and Transversotrematidae have cercariae which develop directly into adults in their definitive host. Cercariae of these flukes actively penetrate into their definitive piscine host (Sommerville and Iqbal, 1991 and Rao and Ganapati, 1967). All other flukes which attain maturity in piscine hosts, reach their definitive host as waiting stage metacercariae

▪ NEMATODES (ROUNDWORMS)

They firmly attach by their buccal capsule to the stomach mucosa. (Paperna, 1964 and Khalil, 1971)

▪ ACANTHOCEPHALA (THORNY-HEADEDWORMS)

Acanthocephala are readily recognised by their evaginable proboscis crowned with several rows of recurved hooks. In the encased larval stage, in tissues, the spiny proboscis is retracted. The worms are sac-like, containing lemnisci connected to the proboscis and genital organs opening posteriorly. The sexes are separate and the male opening is within a membranous bursa. An alimentary canal is absent.The number and arrangement of the hooks on the proboscis is the main criteria for differentiation of species(Kabata, 1985). Pathological effects are localised around the attachment of the adult worm. The extent of damage is proportional to the depth of penetration of the proboscis. It is negligible when parasites are attached to the epithelial mucosa only(Paperna and Zwerner, 1976; McDonough and Gleason, 1981 and Kabata, 1985).

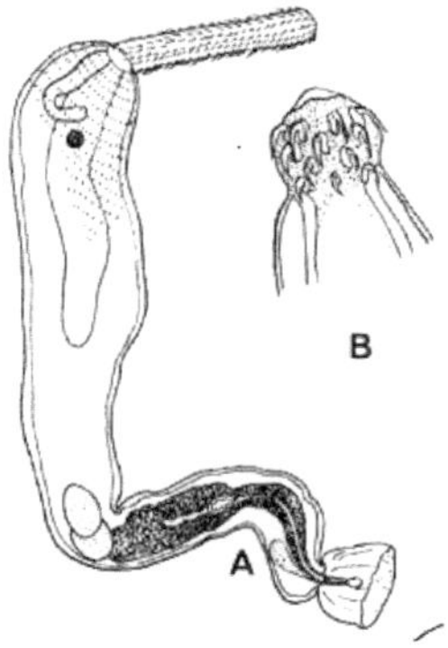

▪ CESTODES (TAPEWORMS)

They have three different body regions: scolex (attachment), neck (buds off proglotids), strobila (remainder).Proglottids formed by the asexual budding of neck. All proglottids develop both male and female sexual organs.

Adult scolex becomes evident at the procercoid stage in the oligochaete intermediate host. The scolex of the adult has 13 to 19 prominent longitudinal ridges and deep furrows that come together at the apex to form an apical ring, a protrusible terminal introvert within the apical ring that forms a deep apical pouch when fully retracted, and a central group of Faserzellen. The terminal introvert may be responsible for attachment in early juvenile stages, but may be supplemented by the longitudinal ridges and furrows later in development. Host tissue appears to be drawn into these furrows that function as weak organs of attachment.(Mohammed H. Ibraheem and John S. Mackiewicz 2006)

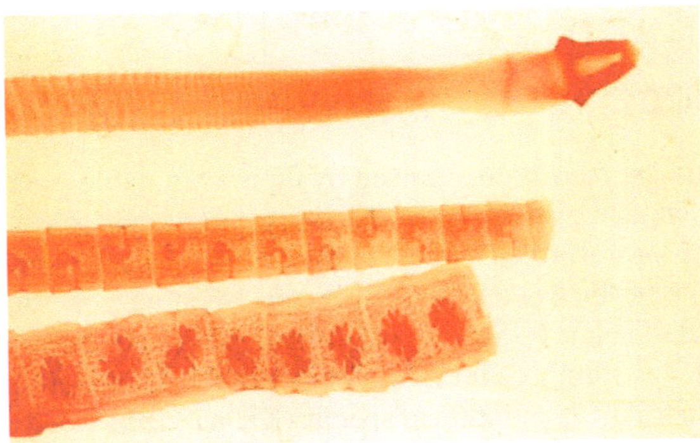

- **CRUSTACEA**
- **Isopoda**

Isopoda are the largest crustaceans found on fish (20–50 mm long).The main part of their body is divided into seven segments, each of which has a leg attached on each side. The legs are armed with sharp sickle-shaped tips which help the isopod attach to its host. These sharp appendages also are responsible for many of the "bites. The next five smaller, partly fused segments are the abdomen, followed by the tail.

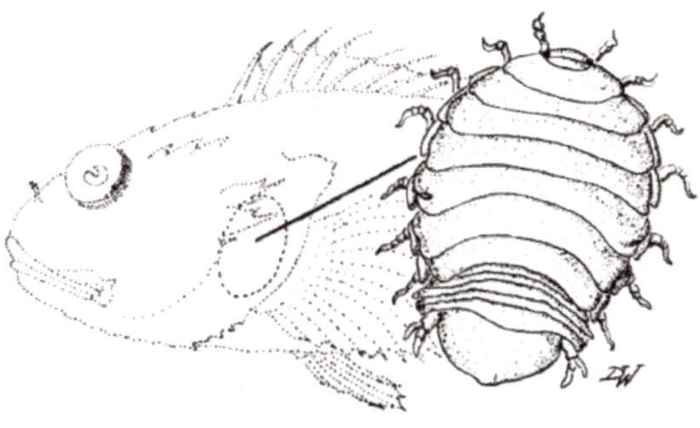

▪ Copepods

Ergasilus (Gill maggots): Only the female is parasitic, and is found on the gills of fish.Females attached to gills produce eggs in two sacs which are attached to the genital segment. Ergasilids attached to gill filaments produce small foci of erosion; Apparently feeding involves excretion of proteolytic enzymes for external digestion(Fryer, 1968a).

Ergasilus

Lernaea

Lernaea (Anchor worm): Head bears 3 horns attached to the concerned organs, while the long tubular neck passes through the body cavity and muscles.The rest of the body consisting of the genital segment, abdomen and egg sacs remains exposed to the outside.The head and major portion of neck were found to be encapsulated in a cyst like formation. The visible parts of the anchored parasite comprised its flexed 6-10mm abdomen with its 4-7mm long, straight or branched, caudal processes.

The cephalic part with its branched processes (antlers, or cephalic horns) as well as the 7-10 mm long tube-like neck linking it to the abdomen, were invariably found to have burrowed deep into the host tissues, causing damage and deformities to the bone and cartilage, destroying the olphactory epithelium, or the tissues behind the eyeball. (Panos Varvarigos, 2007)

Argulus: *Argulus* or fish louse is a lare parasite that attaches to the external surface of the host.The most conspicuous appendage is sucker,a unique modified part of maxillae. Argulus attached to skin & fins by means of two suction discs & sucks the body fluid through sharp proboscis. Damage is caused by the piercing proboscis stylet which enters deep into the skin's dermal layer and secretes lytic and toxic substances resulting in acute haemorrhagic, inflamed wounds (Kabata, 1970, 1985 and Paperna and Zwerner, 1976).

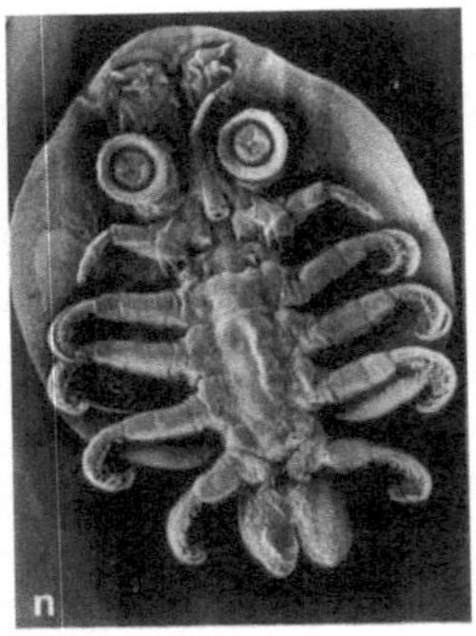

Argulus

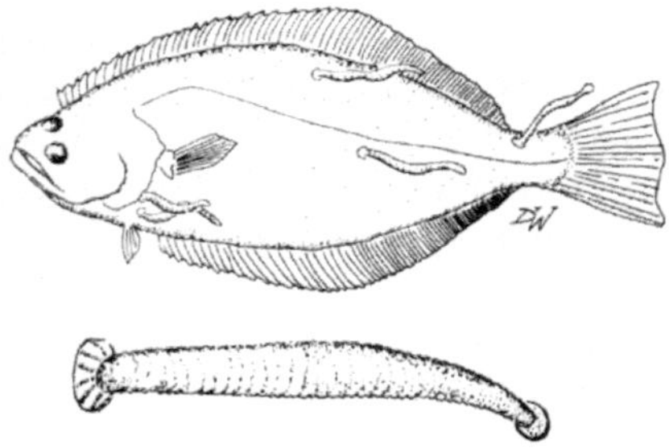

Leeches

▪ LEECHES

Hemophagic (feeding on blood) leeches attach to their hosts and remain there until they become full, at which point they fall off to digest. They all have an anterior (oral) sucker formed from the first six segments of their body, which is used to connect to a host for feeding, and can also release an anesthetic to prevent the host from noticing the leech. They use a combination of mucus and suction (caused by concentric muscles in those six segments) to stay attached and secrete an anti clotting enzyme into the host's blood stream. To further aid feeding, leeches inject histamine into the host's blood vessels. This causes the vessels to dilate, thus increasing the blood flow.

6

Zoonotic Potentiality of Parasites in Cultivable Food Fishes

INTRODUCTION

Zoonoses are diseases and infections which are naturally transmitted between vertebrate animals and humans. The pathogens that cause zoonotic diseases are bacteria, fungi, viruses, parasites or protozoa. Disease transfer occurs via direct or indirect exposure to the pathogen. A pathogen is contracted directly through skin to skin contact of an infected animal or through ingestion of diseased animal products (infected faeces, urine, saliva, blood, milk or other bodily fluids). A bite or sting from a diseased animal or through an infected invertebrate host directly transfers the pathogen to humans as well. Indirect exposure occurs through air, water or soil in which the pathogen can survive for a limited period of time without a host (Acha, *et al.*, 2003).

CLASSIFICATION OF ZOONOSES

The World Health Organization (WHO) Expert Committee on Zoonoses came up with the following classification system of zoonoses based on the type of life cycle of the pathogen.

Direct Zoonoses

Direct zoonoses are transmitted from an infected vertebrate host to another host by direct contact, tomite or mechanical vector. The pathogen does not undergo developmental change or propagation during the transmission. An example is rabies.

Cyclo zoonoses

A cyclozoonoses requires more than one vertebrate host but no invertebrate host is needed.e.g; disease caused by, *Taenia saginata*, *Taenia solium* etc.

Meta zoonoses

A metazoonoses does require an invertebrate host where the pathogen multiples or develops before it can infect a vertebrate host. An example of a metazoonotic disease is bubonic plague.

Sapro zoonoses

Saprozoonoses are diseases transferred through a non-animal reservoir, such as a plant, or through the abiotic environment, such as through water or soil such as *Capillaria hepatica* etc.

SOURCES OF ZOONOTIC DISEASES

As long as you are in the immediate environment of animals or untreated animal products, you are at risk of contracting a disease or infection that can be traced to them. Sometimes toxins or poisons linked to animals are transmitted to us through the injury. Materials such as soil and water that have come in contact with animals, animal by-products or animal waste can also cause disease (Forbes, *et al.*,2005).

How Humans Contract Zoonotic Diseases:

The nose, ears, eyes and mouth serve as easy points of entry for viruses, bacteria and parasites. This means that air; water, food, soil and direct contact with an animal play a part in assisting disease causing organisms or infectious agents to enter your body. Of course, cuts and scrapes in the skin can provide entry points as well.

People At Risk Of Zoonotic Diseases:

Infants and Small Children — their immune systems are not yet fully developed.

Pregnant Women — their immune systems may be weakened.

The Elderly — they may have impaired immune systems.

NEMATODES

Anisakiasis

Anisakiasis is primarily an acute gastrointestinal disease caused by infection with either the herring worm *(Anisakis species) or* the cod worm

(Pseudoterranova decipiens) the larvae reside in the muscles and the visceral organs of marine fish, with the intensity of infection varying among fish species.

Spreading: Human infection occurs by ingesting raw or undercooked fish. The apparent increase in anisakiasis cases in Japan and other developed countries are more.

Gastric Anisakiasis: The larvae usually penetrate the gastric wall causing acute abdominal pain, nausea, and vomiting within a few minutes to several hours.

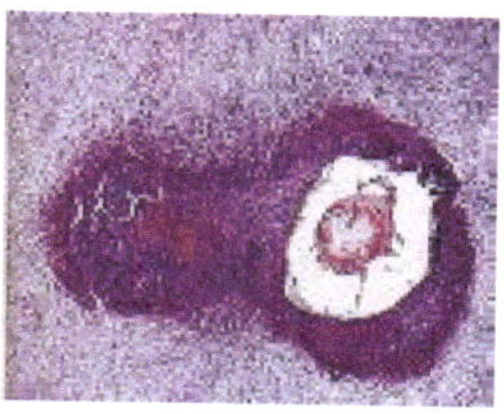

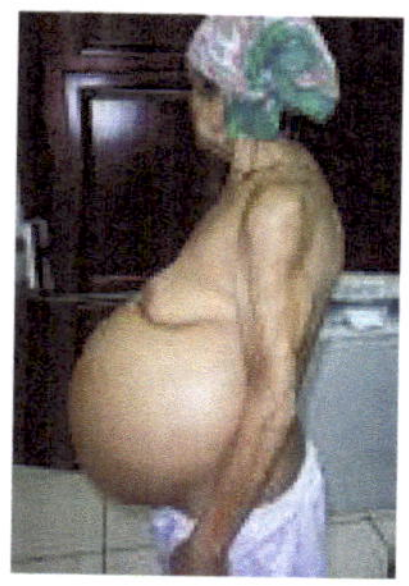

Intestinal Anisakiasis: Although the frequency is low, larvae may invade the intestinal mucosa. Severe local pain occurs and inflammatory reaction often results in reactive intestinal obstruction.

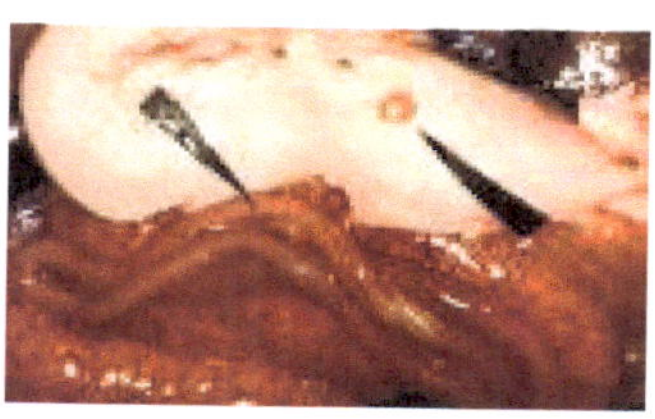

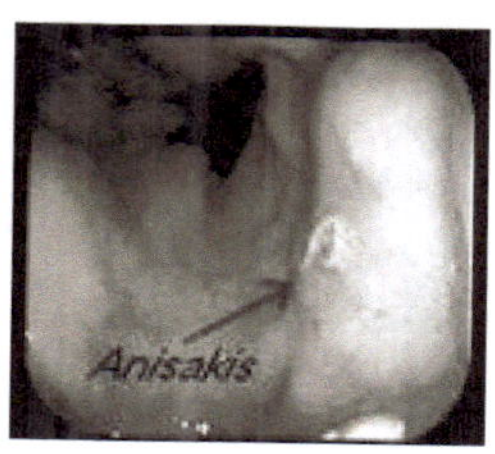

Extragastrointestinal Anisakiasis: Anisakis larvae occasionally penetrate into the peritoneal cavity or other visceral organs to cause eosinophilic granuloma, which is often suspected of being neoplasia.

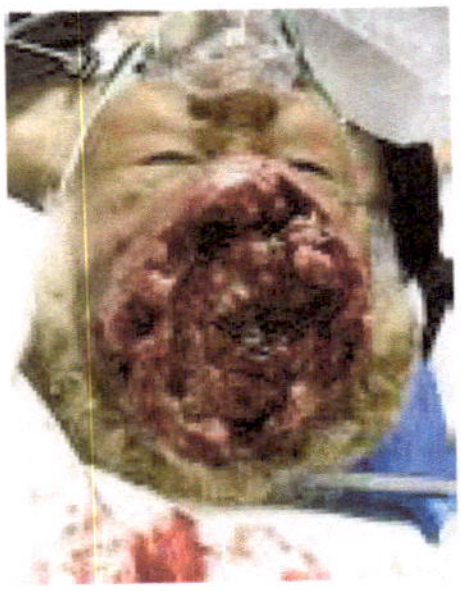
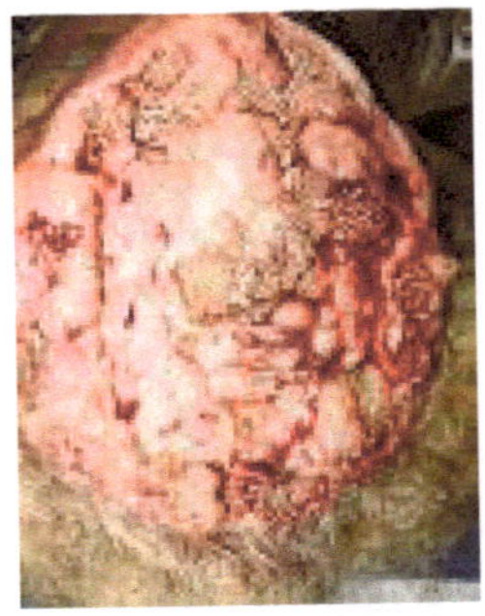

TEST

An antigen-capture ELISA with a reported sensitivity and specificity near 100% appears to be a very appropriate test for serodiagnosis of anisakiasis.

Treatment

The appropriate drug treatment for anisakiasis has not been established. However, albendazole (dosage, 400 or 800 mg for 6–21 days) was shown to be effective (Yukifumi Nawa *et al.* 2004).

Gnathostomiasis

Gnathostomiasis is primarily a disease of the skin caused by migration of the larvae of a *Gnatostoma nematode. Of the 12* distinctive *Gnatostoma species, Gnatostoma spinigerum is widely* distributed in Asia and has been considered to be the only species with human pathogenicity. The recent outbreak of gnathostomiasis in Latin America was caused by *Gnatostoma binucleatum.*

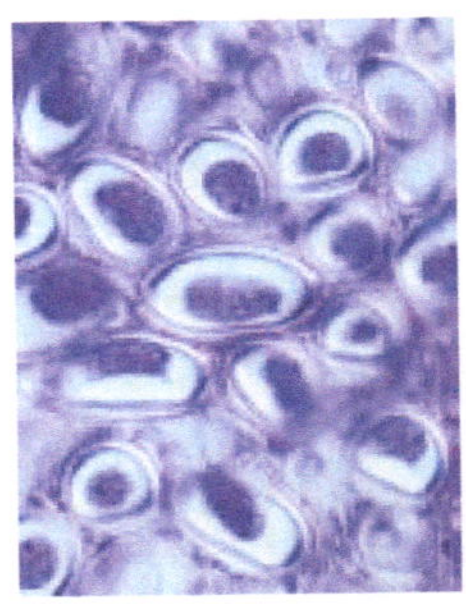

Spreading

Various fish harbour the parasites Snakehead (*G. spinigerum*), Loach (*G. hispidum* and *G. nipponicum*), brook trout *(G. doloresi).*

Effect On Human

Nonspecific prodromal symptoms, such as malaise, fever, urticaria, and nausea, may be observed shortly after ingestion of larvae. Eosinophilia occurs by the time a skin lesion begins to develop. In *G. spinigerum and G. binucleatum infections, the symptoms include the episodic appearance* of migrating erythema, usually on the peripheral parts of the body. In contrast, in *G. hispidum, G. nipponicum, and G. doloresi infections, the larvae tend to migrate into the surface* skin to form a eruption on the trunk. Involvement of the CNS may present as meningoencephalitis, cranial nerve palsy, myelitis, radiculitis, or subarachnoidal bleeding. It is commonly associated with pleocytosis and eosinophilia (15%–90% of the nucleated cells) in the CSF.

Diagnosis

Diagnosis can only be established by identification of the worms by biopsy.

Treatment

Treatment with albendazole at a dosage of 400 mg daily or twice daily for 21 days has been effective against cutaneous gnathostomiasis .A comparable cure rate can be achieved by with a single dose of ivermectin (0.2 mg/kg). For CNS infection, a combination of treatment with albendazole at a dosage of 400 mg twice daily and corticosteroids was successful.

Intestinal Capillariasis

This famous parasitic disease is caused by the nematode *Capillaria philippinensis.* Freshwater or brackish-water fish are the intermediate hosts.

Spreading

Human infection occurs after ingestion of raw fish. It is endemic in restricted areas of The Philippines and Thailand. Sporadic cases have been reported in Japan, Korea, Taiwan, India, and Iran.

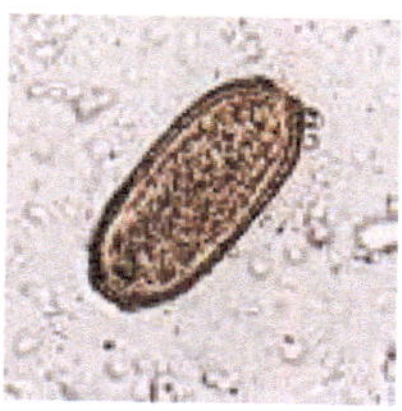

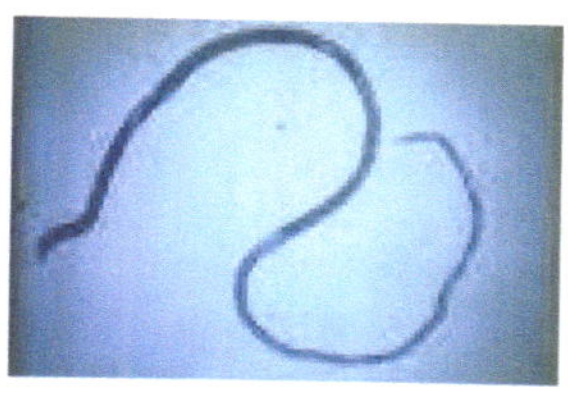

Effect On Human

In the human intestine, adult female worms produce fertilized eggs and larvae.

Patients experience diarrhoea and abdominal pain. If the infection goes untreated, the symptoms get worse with progressive weight loss, weakness, malaise, anorexia, edema, and cachexia and often can result in death.

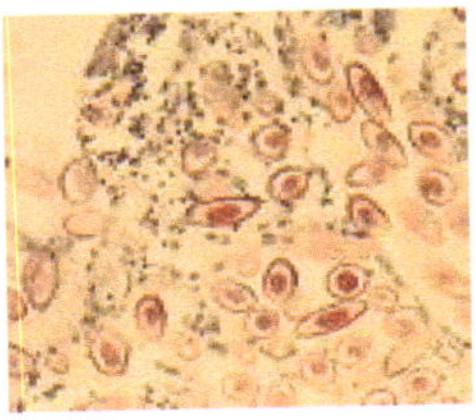

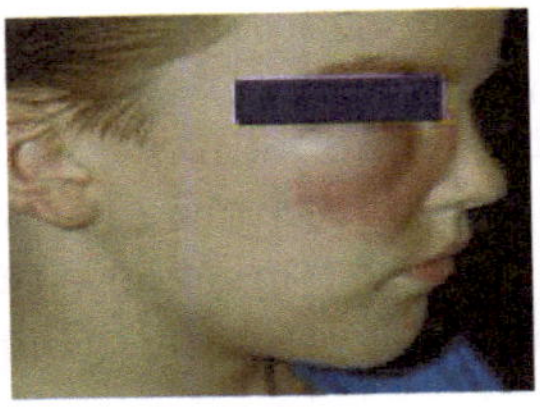

DIAGNOSIS

Diagnosis is made by identifying the characteristic eggs in stool samples.

Treatment

Albendazole at a dosage of 200 mg daily (adult dose) for 10 days is standard.

CESTODES

Diphyllobothriasis

Diphyllobothriasis is an intestinal infection caused by the fish tapeworm *D. latum. Infective larvae (plerocercoid) of D. latum* reside in the muscles of trout, salmon, pike, and sea bass.

Spreading

After being ingested, the plerocercoids attach to the mucosa of the small intestine, where they become adult worms of <"5–10 m in length. The disease is regularly observed in cold climate area.

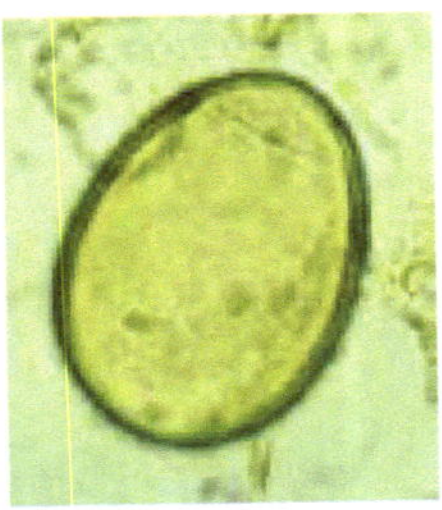

Effect On Human

Infection with *D. latum is mostly oligosymptomatic.* Although diphyllobothriasis is known to cause reduced serum levels of vitamin B_{12} and subsequent anemia.

Diagnosis

The diagnosis is mainly based on the detection of ova or proglottids in the faeces.

Treatment

The treatment of choice is a single dose of praziquantel (10–20 mg/kg).

Sparganosis

Tapeworms of the species *Spirometra* sometimes attack the man as false host, while usually dog, cat and other carnivores are hosts. The two alternate hosts are little crustaceans, fish, frogs and snakes.

Effect On Human

Larva cause lymphocytic and eosinophilic infiltration in the subcutis. If the tapeworms invade the brain, headache, epileptiform attacks and other damages occur. (Yoshikawa , et al.,2002)

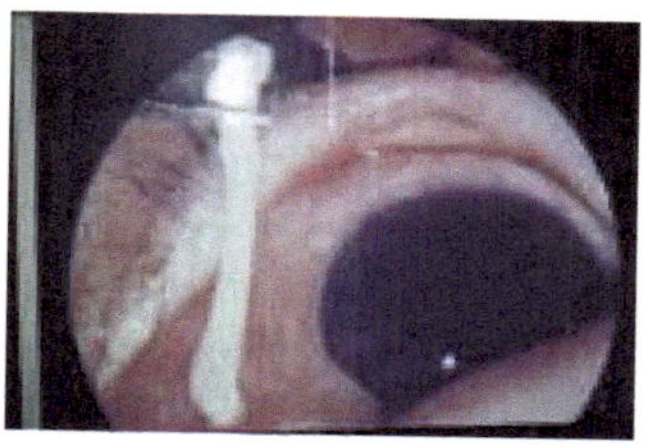

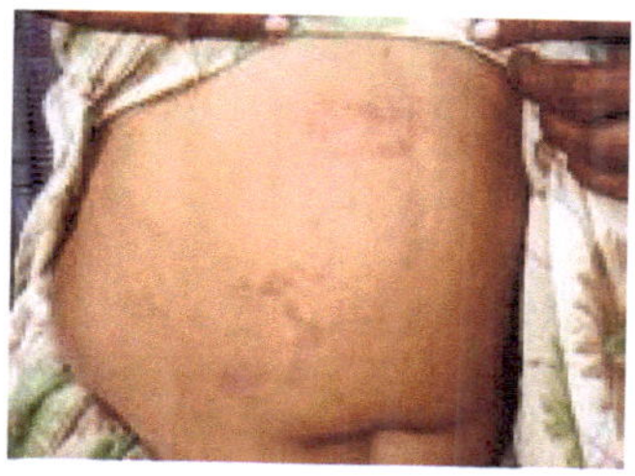

TREMATODES:

Paragonimiasis

Paragonimiasis is caused by infection with *Paragonimus lung* flukes. *Paragonimus westermani is the most common species* in Asia and is the major source of human infection.

Spreading

Infective larvae (metacercariae) reside encysted in freshwater crabs, which may be included in ethnic dishes and ingested. Larvae penetrate the peritoneal cavity and move across the diaphragm into the pleural cavity.

Effect On Human

Abdominal symptoms may be observed during the migratory phase. Finally, the parasites migrate into the lung parenchyma, where they reach maturity and form solid worm cysts. Typical clinical manifestations are fever, chest pain, and chronic cough with hemoptysis (rusty colored sputum).

Diagnosis

Radiographic findings are characterized by infiltrative, nodular, and cavitating lesions. Diagnosis is confirmed by either detection of ova in sputum, stool, or gastric aspirate of specimens or serological test.

Treatment

Regardless of the causative species and the sites affected, patients can be treated with praziquantel at a dosage of 75 mg/ kg (Nawa, *et al.*,2004).

LIVER FLUKE INFECTIONS

Two species of liver flukes, *Clonorchis sinensis and Opistorchis viverrini,* are known to cause hepatobiliary disease.

Spreading

Opisthorchiasis is transmitted as a result of consuming raw or insufficiently cooked freshwater cyprinoid fish (carp).Larvae migrates to bile ducts, where they develop into adult worms.

Effect On Human

Patients infected with only a few parasites are usually asymptomatic. A heavy infection may lead to obstructive jaundice and eventually to liver cirrhosis and cholangiocellular carcinoma, which is a serious public health issue in Thailand.

Diagnosis

The infection is diagnosed by detection of ova in faeces specimens, with the aid of radiographic or ultrasound findings (intraductal echoes and biliary dilatation).

Treatment

Treatment with praziquantel (75 mg/kg per day for 2 days) is highly effective. (Yoshida, 2000)

DIDYMOZOID

Numerous ribbon-like aggregations of eggs of a trematode didymozoid were found embedded in the connective tissue, mostly between the epoxial skeletal muscle fibres, of mature mackerel *Scomber scombrus* (Gibson *et al.* ,1981).

SPREADING

Human are affected by eating fresh, frozen, smoked, or canned mackerel.

Effect On Human

Allergies occur with the presence of this helminth parasite (Thompson, 2000).

MINUTE INTESTINAL FLUKE INFECTIONS

Globally, 70 species of intestinal flukes (trematodes) are known to infect humans. Of those, 31 species belong to the *Heterophyidae family,* and 21 belong to the *Echinostomatidae* family; both are well-known families of fishborne parasites, many of which are distributed widely in Asia and are human pathogens.

Spreading

Infections can occur by ingesting raw fish or snails.

Effect On Human

Minute intestinal flukes are usually harmless, but heavy infections occasionally cause serious gastrointestinal symptoms. Regardless of the causative species, clinical features of intestinal fluke infections are similar.

Diagnosis

Because of the morphological similarities between species, species-specific diagnosis by identification of eggs in stool samples is almost impossible. Differential diagnosis can only be established by distinguishing adult worms purged in stool after treatment.

Treatment

Treatment with a single dose of praziquantel (10–50 mg/kg, depending on the parasite species) is effective for the treatment of minute intestinal fluke Infections. (Waikagul, et al.2005)

METAGONIMUSIS

In Japan, infection with *Metagonimus yokogawai* is still prevalent because people prefer to eat sushi and sushimi of the locally famous freshwater fish "Ayu" *(Plecoglossus altivelis).*

Spreading

Metacercariae infect humans after ingestion of raw or undercooked fish. The flukes then invade the mucosa of the small intestines, causing inflammation and ulceration.

Effect On Human

Patients infected with *M yokogawai* present with mucous, diarrhea, and vague abdominal symptoms.

Detection

M.yokogawai is the most frequently (10%–20%) detected parasite by examination of stool samples for parasite eggs.

Treatment

Treatment with a single dose of praziquantel is effective, (N.D. Thu *et al.* ,2007).

HETEROPHYIASIS

H heterophyes is the most common of the 10 species that compose the genus Heterophyes.

Spreading

The second intermediate hosts—brackish or freshwater fish. In the second intermediate hosts, the cercariae are transformed into metacercariae, which infect humans upon ingestion of raw or undercooked fish.

Effect On Human

H heterophyes are observed in the human intestine, jejunum, and ileum. In humans, the flukes attach to the small bowel and cause shallow ulcers, mild inflammation, and/or superficial necrosis. Clinical presentation includes diarrhea, dyspepsia, and intestinal colic. Because of their small size, the eggs, and sometimes the adult flukes, enter blood vessels and embolize to the brain, producing symptoms similar to cerebral haemorrhage. Eggs may also enter the mesenteric lymphatics and travel to the heart, causing myocarditis, chronic congestive heart failure, and death (Choon Shil Lee, 2009).

PROTOZOAN DISEAES

Amebiasis

The cysts of *Entamoeba histolytica* are transmitted by contaminated water or food onto people. The disease is spread worldwide, about 10 % of the world's population carries amoeba. The symptoms reach from moderate to mucosal-bloody diarrhea. The most frequent complications are metastases in the liver and/or lungs including abscesses. The abscesses in the liver tend to penetrate into the pleural cavity.

Giardia (*Giardia Lamblia)*

Giardia is one of the most common intestinal parasites in the United States. It is primarily a waterborne parasite. Most infections occur due to the ingestion of contaminated water. The primary symptom of a *giardia* infection is diarrhea, but other symptoms such as flatulence, stomach cramps and nausea may be present.

Coccidia (*Cryptosporidium*)

Coccidians are single-celled parasites that cause diarrhea. Infection occurs due to the accidental ingestion of food or water contaminated with *Crytosporidium.* Symptoms are watery diarrhea, weight loss, fever, abdominal pain, nausea and vomiting. No real treatments exist for this parasite.

CONCLUSION

Marine fish for raw consumption should be frozen at less than 20 ^{0}C for more than 24 h. Many zoonotic parasites use water as a vehicle for transmission to their definitive or intermediate hosts. Implementation of on-farm control measures to eliminate the contamination of agricultural water with human faeces is a key factor in the control of many zoonotic parasitic infections. (Heejeong Youn, 2009

7

Effect of Parasites on Fish Behaviour

For most people in their right mind, parasites are disgusting little creatures. The mention of them evokes images of tiny crawlers entering the body of innocent victims by devious means, encrusting themselves within various organs (often with very unpleasant effects), living off their host's tissues, and then exiting the body, sometimes through scatological avenues, leaving behind hosts that are incapacitated, if not mere shells of their former selves (as is literally the case for some insects). The true facts are not always that gruesome (after all, parasites that have to spend a long time inside a host do not benefit if this host kicks the bucket, because then they die as well) but often enough parasitic reality is not pretty. Ask biology students about their parasitology course, and they will tell you that attending class is like visiting a little shop of horrors. I, for one, came out of my parasitology course with newly developed phobias (mercifully temporary) of such innocuous activities as eating meat and bathing in streams – they represent ways of transmission for some parasites. But students will also tell you that parasites are fascinating because of the complexity of their life cycles and the intricate ways in which they complete these cycles. So fascinating, in fact, that many students eventually become parasitologists and devote their entire lives to studying the little critters. Most of these people work with parasites of livestock or humans, but a few with an equal taste for ethology and ichthyology pursue the more esoteric questions that occupy this page: how fishes avoid getting parasitized, and how fishes alter their behaviour once they get parasitized (Barber *et al.*, 2000)

HOW FISHES AVOID GETTING PARASITIZED

This question has not been studied very much, and the few experiments I am aware of have all dealt with bloodsucking ectoparasites. These small invertebrates can be free living, but when they perceive a passing shadow or some water disturbance, they jump. With luck, they land onto the fish that created the disturbance. They fixate themselves onto the skin or the gills and then they start their dirty work, sucking blood like miniature versions of lampreys. Because of this sit-and-wait strategy, ectoparasites resemble predators, and we would expect that any means of avoiding predators (shunning dangerous areas, shoaling more, moving less) should also work against these parasites. This, in essence, is what the studies have revealed. Almost all of this work was carried out in the laboratory of Gerry FitzGerald, using populations of three-spined and blackspotted sticklebacks that were plagued by the crustacean ectoparasite *Argulus canadensis* (the Canadian fishlouse). Free-living parasites, along with infected and non-infected sticklebacks, could be captured in the field and brought back to the lab.**1** A doctoral student, Robert Poulin, noted that when one half of his tanks was decorated with plastic plants while the other was left open, the fishlice preferred to stay near the bottom in the "vegetated" area. Sticklebacks also preferred to swim near the bottom in vegetated areas but only when parasites were absent (Tripathi, 1956).

When free-living fishlice were present, the sticklebacks' choice of habitat changed: they spent more time near the surface and in open areas, presumably to avoid the more infested sectors of the tank.**2** Poulin then investigated another strategy for avoiding these parasites: shoaling more. He released 30 sticklebacks in circular wading pools that might or might not contain free-living fishlice. He reported that, as compared to the absence of parasites, fish in the presence of lice spent more time shoaling and formed larger groups. The benefit from bigger shoals derived from a higher probability that the free-living parasites would be detected and eaten by shoalmates – fishlice cannot infect a fish from the inside and can therefore be consumed with impunity. Another benefit was a dilution of the risk of being targeted for an attack by any given parasite.**3** Poulin then switched to another system (and another lab) and made the following observation. Within experimental tanks, brook trout fry that ended up getting infected more quickly by ectoparasitic copepods were also those that moved more within their environment.

This result suggested that reduced movement is another possible strategy for avoiding ectoparasites. **4** Later, back in FitzGerald's lab, Lee Dugatkin wondered if sticklebacks would avoidthe company of parasitized conspecifics, given that fishlice can jump from fish to fish. He divided a long aquarium into three sections with transparent partitions. He put 10 healthy fish on one side and 3 healthy fish together with 7 infected ones on the other. He then repeatedly put single fish in the middle and recorded which group they approached most often. By and large the sticklebacks spent more time close to the healthy school, probably because the infected fish on the other side displayed odd behaviours such as swimming erratically and rising to the surface. **5** (The avoidance, therefore, probably was not to external signs of parasitism *per se*, but to conspecifics that behaved abnormally. Nonetheless, the end result is that parasite transmission is reduced). Even in the case of ectoparasites that cannot be transmitted directly from fish to fish, it may pay to avoid the company of parasitized shoalmates. This is because the abnormal behaviour or appearance of infected fish may draw the unwanted attention of predators that could then endanger the whole shoal. One workable system here is the association between the trematode worm *Crassiphilia bulboglossa* and its host, the banded killifish *Fundulus diaphanus*. This trematode encysts itself in the muscles of its host, and this cyst forms a black spot that can be seen externally through the skin. Jens Krause showed that, in choice tests, banded killifish preferred to associate themselves with completely unparasitized shoals rather than completely parasitized ones (surprisingly however, Krause's killies did not prefer unparasitized shoals over partially parasitized ones).**6** The same experiment also revealed that, in this case, it is the black spot caused by the worm's presence that betrays infection to the eyes of shoalmates. When members of a shoal were injected subcutaneously with a speck of black ink, they still displayed the normal behaviour of healthy fish but were nevertheless shunned by their shoalmates in choice (Srivastava, 1975) .

How fishes alter their behaviour once parasitized

This brings us to the second question, that of altered behaviour in parasitized fish. The underlying theme here is that often fishes are not the only host in a parasite's life. An endoparasite's life cycle often starts with the first larval stage entering the body of an aquatic invertebrate such as a snail. From the snail a new larval form is transmitted to a fish, either through ingestion of the snail or directly through water.

Once inside the fish, another larval stage of the parasite encysts itself. Then the fish gets eaten by another fish or by a bird, the final host, inside of which the adult parasite develops. The eggs of the adult parasite are then passed on in the predator's faeces, hopefully over water in the case of a bird, and from there infection of a snail can start the cycle anew. So, once inside a fish, it is in the parasite's interest to somehow alter the behaviour or appearance of this fish in a way that increases its chance of being captured and eaten by the next host in the cycle. Most studies have looked for evidence of this so-called "manipulation hypothesis". The most convincing results to date have come from a study by Kevin Lafferty and Kimo Morris, working with the Pacific (California) killifish *Fundulus parvipinnis* and its two trematode parasites, *Euhaplorchis californiensis* and *Renicola buchanani* The researchers put wild-caught killifish in a 150-liter tank, selected focus individuals, measured their behaviour, and caught them again in order to dissect them and count the number of parasites inside. They found correlations between the level of infection and the frequency of conspicuous behaviours such as flashing, contorting, shimmying, and coming to the surface. The heavier the infestation, the more common the odd behaviours were. Lafferty and Morris then tested the idea that such behaviours could make the fish more susceptible to predation. First, they captured killies from two different populations, one known to be parasitized and the other never parasitized. Then they placed known numbers of these fish in outdoor pens accessible to various species of wild herons, gulls, and fish-eating ducks. After 20 days, only 4 out of 53 unparasitized fish had disappeared (that is only 8%), as compared to 51 out of 95 parasitized ones (54%). Disappearance was caused by predation and not by other factors such as disease or escape from the pens: the study included a control experiment with enclosures covered by netting that prevented bird attacks, and almost no fish from either population disappeared from these pens. Lafferty and Morris concluded that parasitized fish are more susceptible to avian predators, to the benefit of their parasites.**7** In that study, how the parasites altered the fish's behaviour remained a mystery. (The fact that one of the trematode, *Euhaplorchis*, was found in the brain, suggests that the production of some crucial neurotransmitters may have been manipulated by the parasite). Other studies have explored parasite-host systems that are more telling in that regard. For example, the common dace *Leuciscus leuciscus*, when heavily parasitized by the eyefluke *Diplostomum spathaceum*, has been found to spend more time in the top 10 cm of an aquarium, possibly because the eyesight of the fish is impaired by the parasite, causing a preference for the better-lit surface waters.**8** The ultimate hosts of this parasite, gulls, are known to be more successful at catching fish that are closer to the surface. The same parasite can also affect the eyesight of rainbow trout, reducing the fish's ability to match its colour with that of the environment, thus making them more visible to predators, and reducing also their reaction to moving objects – some of which might be predators.**9** Another commonly studied system is that of sticklebacks and the cestode *Schistocephalus solidus*.

These parasites live in the peritoneal cavity of sticklebacks in such great size and numbers that they sometimes cause a visible distension of the abdomen. The parasites divert for their own purposes much of the energy assimilated by the fish. Afflicted fish are therefore hungrier, and we know that hungrier fish are more willing to take risks in order to keep on foraging. Not surprisingly therefore, several studies have shown that parasitized sticklebacks, after being scared by the strike of a model heron bill, flee over shorter distances, remain motionless for shorter periods, and come back to a food patch sooner than healthy individuals.**10** They are also willing to risk looking for food at the bottom of tubes that are located fairly close to a big predatory cichlid (behind a clear partition), something that healthy fish will not do.**11** In a similar vein, parasitized nine-spined sticklebacks that perform aquatic surface respiration in hypoxic waters (see the page: Oxygen and fish behaviour) come back to the surface sooner than nonparasitized fish after being frightened by the passage of a model kingfisher over the aquarium.**12** Infected fish may have increased energetic demands, requiring more oxygen. Hungrier fish are also known to form shoals that are smaller, less compact, and less cohesive, attributes that favour foraging but make the shoal more vulnerable to predation.

In addition, hungry fish spend more time at the frontal periphery of the shoal where they can be the first to find food, even though they may also be the first to encounter predators. Do parasitized fish show the same tendencies? Many studies have revealed that they do. Sticklebacks with cestodes, European minnows with helminthes, fathead minnows with flukes, as well as killifish with trematodes, leave more room between themselves and their neighbours, and stay at the front or periphery of the shoal – or leave the shoal altogether – more often than their healthier counterparts.**13** In general, the method used in these studies was to set up a shoal, select a focus fish that was obviously parasitized (as betrayed by abdominal distension or by black spots on the body), measure its social behaviour, and compare it to that of a healthy individual within the same shoal. That method is not as elegant as the one used by Lafferty and Morris. As described above, these two researchers identified some fish by external marks unrelated to parasitism, measured their behaviour, and dissected them later to learn of their parasite load. This is an example of what we call a "blind" experiment because the observers did not know in advance what category (heavily parasitized, slightly parasitized, unparasitized) the focal fish would belong to. Therefore, they could not be biased while scoring the fish's behaviour. Observer bias – an unconscious tendency to measure ambiguous behaviour in a way that will make the results fit the expectation – is a constant danger in ethological studies. Researchers must compensate (but not overcompensate!) by trying to stay conscious of this danger or by using blind protocols (Lafferty, *et al.* 1996).

Parasitism and mate choice

It may sometimes fit the ethologist's purpose not to know in advance the level of parasitism in fish, but the same cannot be said of the fish themselves. Females looking for a mate, for example, probably want to know if the prospective father of their young will have the stamina to care for the eggs and fry (in the case of parental species), and stamina is generally curtailed by parasitism. Females would also prefer males that could pass on to their offspring whatever heritable trait they may possess for resistance to parasites. Females would therefore be looking for a body feature that could reveal the male's infection level.**14** By choosing to mate with males that have such features, with a preference for those males that show marks of health, females could create a selection pressure for the existence of sexual ornaments that reliably divulge the male's level of parasitism. This hypothesis was first published in 1982 by William Hamilton and Marlene Zuk,**15** originally in the context of bright colors in birds, but tested a few times with fishes since then. The most striking of these tests was carried out by Manfred Milinski and Theo Bakker at the University of Bern. They started by placing two male three-spined sticklebacks next to a tank containing a female that was ready to spawn, and noting how much time the female spent near each male. They found that she usually spent more time close to the male with the brightest red color on his belly. That this preference was based on color and not on some other correlated characteristic (such as courting intensity, for example) was elegantly demonstrated by offering the same choice of two males under green light, which rendered the red color uniformly black, and finding that the female then chose at random. Milinski and Bakker next offered a choice between males that had never been parasitized versus males that had been recently infected by a ciliate, *Ichthyophthirius multifiliis*, a parasite that impairs body condition and reduces the intensity of body colors. By the time of the tests, the external white spots caused by the ciliates had dropped off the fish, but the red coloration of the formerly parasitized males was still duller than that of their unparasitized conspecifics. As expected, the female sticklebacks spurned the dull males under normal lighting, but not under green lighting. Conclusion: when it is visible, the red coloration of breeding male sticklebacks is perceived by females as a badge of health, including resistance to parasites.**16** Similar results and conclusion have been obtained with guppies. When infected by external parasites, male guppies lose brightness in their orange spots, and they also court less assiduously. Because of this, females shun them as prospective mates.**17** Avoidance of this kind can work both ways, and in some species the tables are turned on the females. In pipefishes for example, if females show external signs of parasitism such as black spots – either natural ones or artificial ones that have been created by injecting ink underneath the skin – males spend less time next to them.

Other effects of parasites

18 Parasite infection can be correlated with behaviours that do not seem to give rise to increased predation risk. Some of these effects suggest a debilitating effect of parasites, but others are harder to interpret. For example: Salmon harbouring helminth parasites seem to orient less well during their seaward migration.**19** Infected minnows are less well synchronized with their shoalmates.**20** Upland bully *Gobiomorphus breviceps* and guppy females infected with various parasites have problems discriminating between males of different quality and size.**21** Parental male three-spot damselfish, *Stegastes planifrons*, experience more egg loss when they are parasitized, though it is unclear if this is due to a lesser ability to protect the nest from predators or from parasite-induced hunger which leads to egg cannibalism by the father.**22** Sticklebacks blighted with *Schistocephalus* (them again!) generally prefer to eat smaller prey.**23** Perhaps the infected sticklebacks cannot compete with other fish for the larger food items, or maybe the overabundant internal parasites constrict the stickleback's stomach so much so that only small prey can be consumed. On the other hand, though sticklebacks parasitized with *Schistocephalus* consume smaller prey items, when they have an ample food supply and no competition they seem to eat a lot of those small preys, as they end up growing *faster* than non-parasitized fish, even when the weight of the parasites is subtracted. **24** It is unclear whether this is a response of the host fish to try and better marshal defenses against the parasite, or a manipulation of the host's physiology by the parasite to make it more visible to predators. Faster growth leads to bigger body size, and bigger fish are more visible to predators (Shirakashi, *et al.* 2001).

8

Diseases and Parasites of Sports Fishes

Occasionally anglers catch fish that show signs of infection or parasitism. Is the fish safe to eat? The usual and safest response to this question is, "When in doubt - don't." However, very few fish diseases can be transferred to humans. Almost all fish are safe to eat when thoroughly cooked, smoked or frozen.

Fishes are a valuable resources for hill streams and temperate region people. Not only do they provide important recreation, but they are also a valuable supplement to the family food budget. The anticipation of any fisherman, whether he be a young boy with a cane pole and a can of worms or an "old pro," is a quality fish for the table. It is the unwritten law of any true sportsman to utilize the fish he catches. The purpose of this publication is to help you, the angler, determine if the fish you've caught is worthy of your skillet. Remember, a sick fish will not take the hook (Taraschewski, 2000).

PARASITIC DISEASES AND OTHER CAUSES OF ABNORMALITIES

For convenience, we have listed groups of parasitic diseases and other causes of abnormalities sometimes seen in the fisherman's catch.

1. *Viruses and Bacteria:* These microorganisms cannot be seen with the naked eye. They cause many diseases in fish. Symptoms include swollen, fluid-filled body cavity, bulging eyes, bloody fins, bloody spots on the skin or base of scales, sores, and lesions, etc.

2. *Fungi:* Fungi are small, strand-like parasites. They normally do not infect healthy fish. They usually grow on dead tissue or infected wounds. Most of the fungi on fish have a patchy, gray-white, cotton-like appearance.

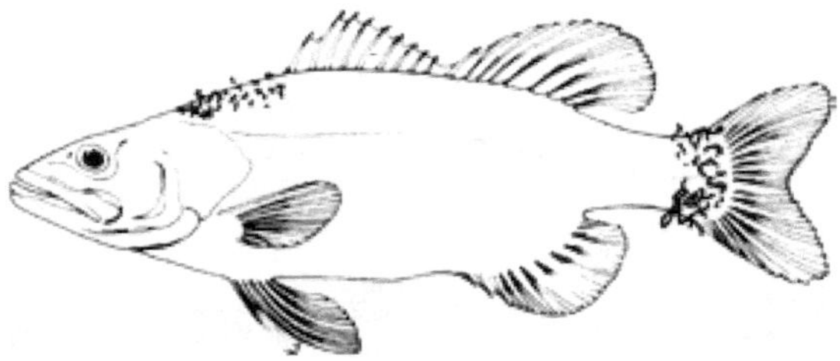

3. *Protozoa:* These are microscopic, single-cell animals. They can be found on the gills, body surface or embedded in the flesh. There are many different protozoans, and they cause a variety of fish diseases. Fisherman usually observe the damage they cause but rarely see the microorganism without the aid of a microscope.

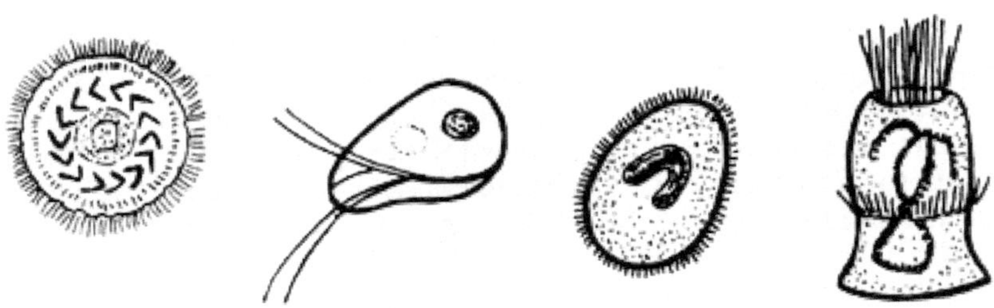

4. *Trematode or Fluke:* There are two groups of flukes. Monogenetic flukes live on the external body surface and multiply on the same host. Digenetic flukes are internal fish parasites and require two or more hosts to complete their life cycle. Other hosts may include snails, clams, birds or other fish - but not man. Trematodes can be found in cysts in the flesh or other internal organs. They also can live in the eye, blood, gills and other parts of the fish. Trematodes are rarely observed by fisherman.

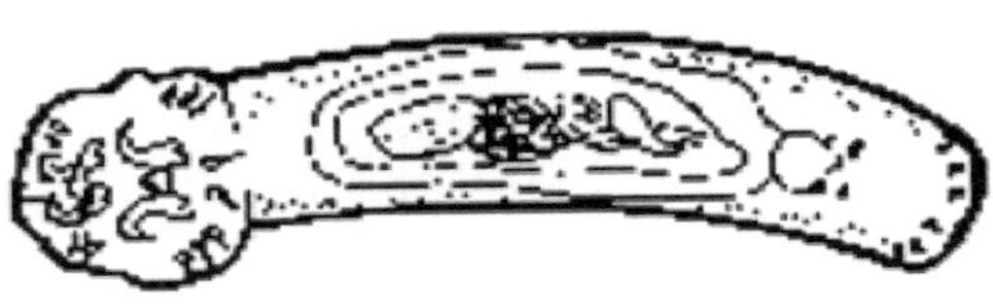

5. *Cestode or Tapeworm:* These parasites are commonly observed by fishermen when cleaning their catch. Larval tapeworms form cysts on or in the internal organs or in the body cavity. Adults are white and worm-like and are found in the intestines.

6. *Acanthocephalan or Spiny-Headed Worm:* These parasites are rarely seen by fishermen. They live in the intestines of fish.

7. *Nematode or Round Worm:* Nematodes are very common parasites of ski. The larvae may be found in cysts or coiled in or on the internal organs. Adults are usually found in the intestines. Some are found coiled under the skin.

8. *Leeches:* These parasites are external, blood-feeding animals. They leave circular wounds on the fish.

9. *Copepods:* These small crustaceans can be embedded in flesh, gills or mouth. Some move freely over the body surface.

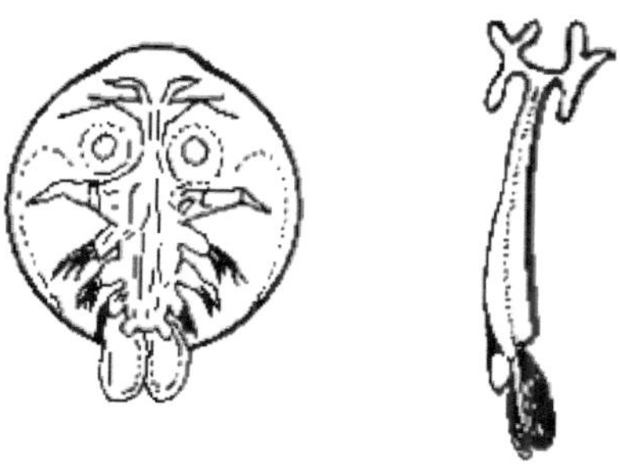

10. *Tumors:* As on other animals, tumors occasionally occur on fish.
11. *Physical Injury:* Predators, such as birds and other fish, can cause wounds and other malformations (Sineszko, et al., 1970).

External Observation:

Visual Signs	Comments
Lesions, sores, hemorrhages, fish pop-eyed, blood under scales, or loss 	Various species of *bacteria* can infect fish. Symptoms of scales vary depending on bacteria and fish species. Bacterial infections are usually the result of a stress on the fish or infection of a wound. An infected fish is edible. Trim away infected flesh.
Cotton-like, white-tan-gray fuzzy growth on body or fins. 	*Fungus* infection of wound or lesion. An infected fish is edible. Trim away infected flesh.
Small, pinhead-size white spots on the skin of catfish and sometimes excessive mucus (slime) production. 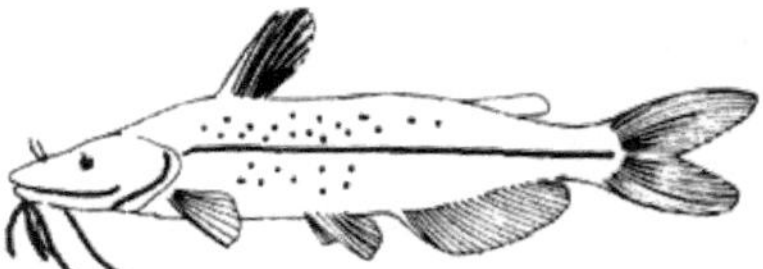	*Ich (Ichthyophthirius)* is a common protozoan parasite of catfish. It occurs on the skin and gills of catfish and some other fish species. An infected fish is edible. Clean and prepare as usual.
Small, black-to-purple spots under the skin or in the flesh of scale fish. 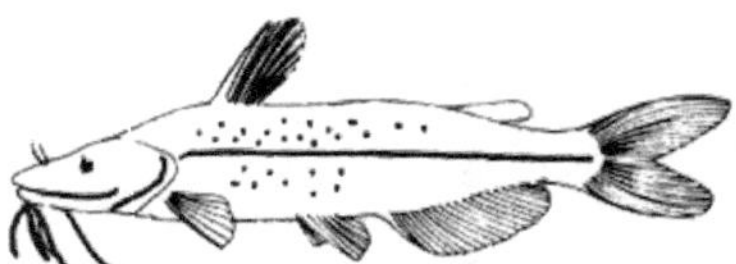	*Black Spot* is one of the more frequent parasites observed by fishermen. It is caused by larval flukes encysting under the skin or in the flesh. An infected fish is edible. Clean and prepare as usual.

Eye opaque or deformed.

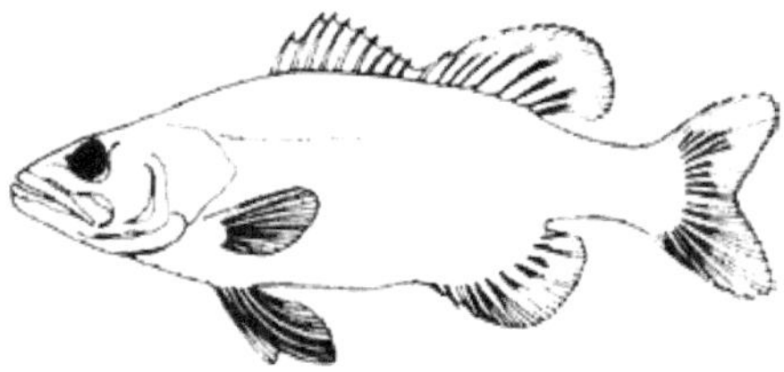

Eye Flukes live in the fluid of the eye. Although they cannot be seen by the fishermen, they eventually cause blindness in the fish. An infected fish is edible. Clean and prepare as usual.

Gills swollen and pink

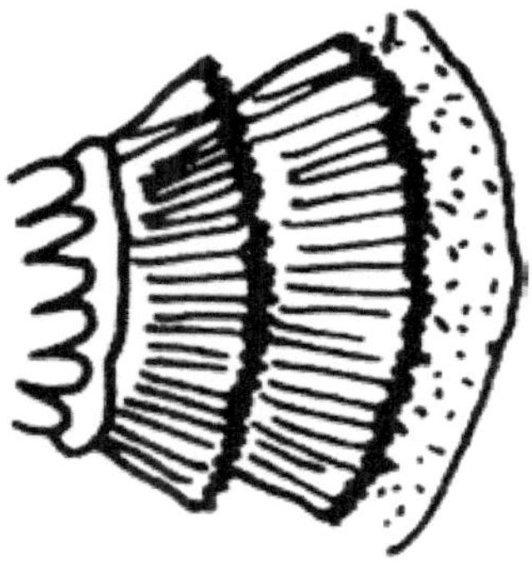

There are a number of *gill parasites* infecting fish. They are microscopic and only the damage they cause is observed by the fisherman. Some of these parasites are *gill flukes* and a number of *Protozoan* species. Chemicals in the water can also irritate the gills and present the same symptoms. Unless chemicals are suspected, the fish is edible. Clean and prepare as usual.

Thread-like red worm extending from the anus.

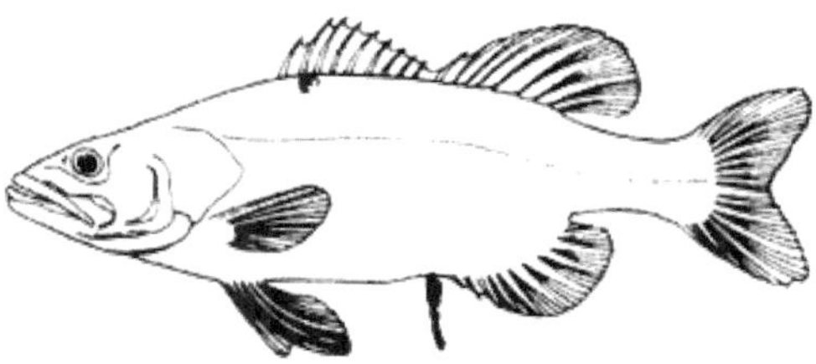

Roundworms can be found throughout the intestines. They sometimes can be seen extending from the anus. An infected fish is edible. Clean and prepare as usual.

Worm-like animal attached to the body, head, fins, or gills.

Leeches are blood-sucking animals that leave a circular wound after they have dropped off the fish. An infected fish is edible. Clean and prepare as usual.

Small, red pustual with red thread-like body protruding from wound at the base of scale or on or near the base of fins.

Anchor Worm is an appropriate name for this parasitic copepod. It buries its anchor-shaped head into the flesh and allows its body to hang free of the wound. An infected fish is edible. Clean and prepare as usual.

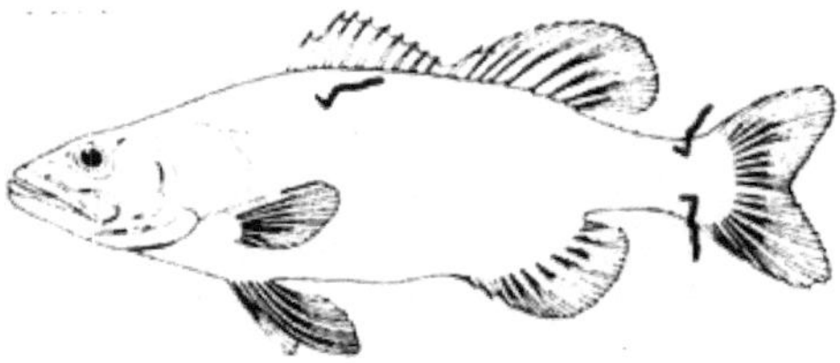

Small, bloody areas at the base or under the scales of a fish.

Fish Louse are microscopic copepods rarely seen by fishermen. They feed on the blood by piercing. the skin. The bites can become infected. An infected fish is edible. Clean and prepare as usual. (Schaperclaus, 1933).

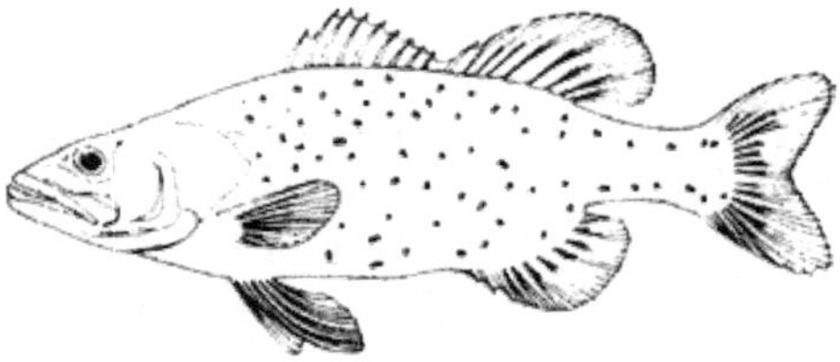

Internal

White worms in the intestine.

Observation:

Adult *flukes, tapeworms, roundworms* or *spiny-headed worms* are commonly observed in the intestines of fish. The fish is edible. Clean and prepare as usual.

Large, white flat worm in body cavity.

Tapeworm found in the body cavity of carp suckers, minnows and some other fish. The fish is edible. Clean and prepare as usual.

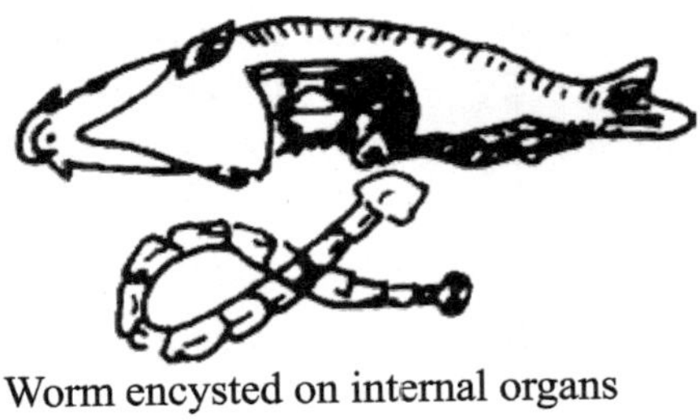

Worm encysted on internal organs coiled like a watch spring.

Larval Roundworm. The fish is edible. Clean and prepare as usual.

Small cysts on internal organs.

Larval Flukes. The fish is edible. Clean and prepare as usual.

Small, white thread-like worms on or in internal organs.

Larval Tapeworms. Often found in the ovaries. The fish is edible. Clean and prepare as usual.

Small, white-to-orange worm in body cavity or attached to intestines.

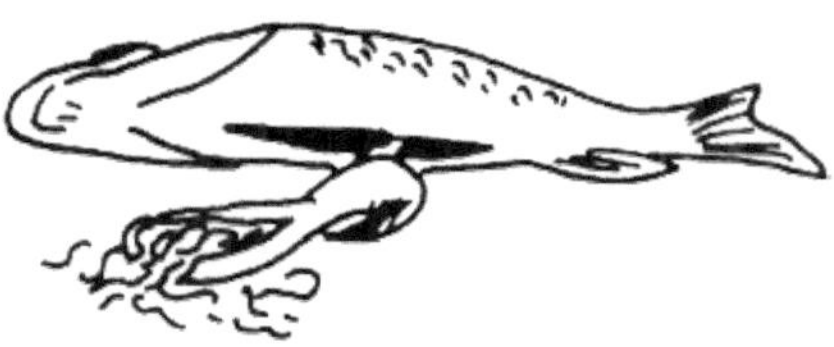

Spiny-headed worm usually lives inside intestines. Sometimes they are found lying in the body cavity with their heads buried in the intestines. The fish is edible. Clean and prepare as usual (Schaperclaus, 1933).

9

Parasites of Exotic Fishes

INTRODUCTION

A parasite is an organism which exploits another organism for the purpose of staying alive. Some parasitic relationships are harmless, while in other cases a parasite can damage or even kill its host. The word is borrowed from the Greek word "parasitos", meaning "one who eats at the table of someone else." In case of exotic carps parasites causes a great loss by reducing economy. Some, cannot live once their host dies, while others can continue thriving on dead hosts until their nutrients are consumed. Parasites live inside a host are called endoparasites or internal parasites. Ectoparasites live outside the host, and are generally more capable of switching hosts. A parasite preys on other parasites, is called epiparasite (Anderson, 1974).

Silver Carp

The silver carp *(Hypophthalmichthys molitrix)* is a species of freshwater cyprinid fish, a variety of Asian carp native to north and northeast Asia. It is cultivated in China. Pound for pound, more silver carp are produced worldwide in aquaculture than any other species. Silver carp are usually farmed in polyculture with other Asian major carps, or sometimes Indian major carps or other species. It has been introduced to, or spread into via connected waterways, at least 88 countries around the world.Introduced in india in 1959 at the pond culture division of CIFRI. In India in 1963 breed sucessfully by hypophysation. The silver carp is a filter feeder, and possesses a remarkably specialized filtration apparatus capable of filtering particles as small as 4 μm. The gill rakers are fused into a sponge-like filter, and an epibranchial organ secretes mucus which assists in trapping small particles. A strong buccal pump forces water through this filter. Silver carp, like all Hypophthalmichthys species, have no stomachs; they are thought to feed more or less constantly. Silver carp are thought to feed largely on phytoplankton; they also consume zooplankton and detritus. It is used in controlling water quality, especially control of noxious cyanobacteria (blue-green algae), microcystis .

Common Carp

The Common carp *(Cyprinus carpio)* is a widespread freshwater fish of eutrophic waters in lakes and large rivers in Europe and Asia. The wild populations are considered vulnerable to extinction, but the species has also been domesticated and introduced into environments worldwide, and is often considered an invasive species. It gives its name to the carp family Cyprinidae. Common carp prefer large bodies of slow or standing water and soft, vegetative sediments. A schooling fish, they prefer to be in groups of 5 or more. They naturally live in a temperate climate in fresh or slightly brackish water with a pH of 6.5–9.0 and salinity up to about 5‰ and temperatures of 3 to 35°C. The ideal temperature is 23 to 30 °C, with spawning beginning at 17–18°C; They prefer to scavenge the bottom for insects, crustaceans (including zooplankton), and benthic worms. In Victoria, Australia, Common carp has been declared as noxious fish species .

Grass Carp

The Grass Carp *(Ctenopharyngodon idella)* is a herbivorous, freshwater fish species of family Cyprinidae, and the only species of the genus Ctenopharyngodon. It is cultivated in China for food but was introduced in Europe and the United States for aquatic weed control Introduced in India at the pond culture division of CIFRI (Orissa) in 1959. Breed successfully by hypophysation in 1963. In the United States, it is known as White Amur, derives from the Amur River, where the species is probably native. It is stocked in many states as an effective bio control for undesirable aquatic vegetation. This species occurs in lakes, ponds, pools and backwaters of large rivers, preferring large, slow-flowing or standing water bodies with vegetation. In the wild, grass carp spawn in fast-moving rivers, and their eggs, which are slightly heavier than water, develop while drifting downstream, and kept in suspension by turbulence. The eggs are thought to die if they sink to the bottom. Adults of the species feed primarily on aquatic plants. They feed on higher aquatic plants and submerged terrestrial vegetation, but may also take detritus, insects, and other invertebrates (Kabata, 1985).

ABUNDANCE OF PARASITES IN GENERAL

H. molitrix: abundance was highest in January and lowest in November *C. carpio var.* abundance was highest in June and lowest in April. *C. idellus:* abundance was highest in February and May and lowest in December. *C. carpio var.* communis abundance was highest in February and lowest in October. *P. gonionotus:* abundance was highest in January and lowest

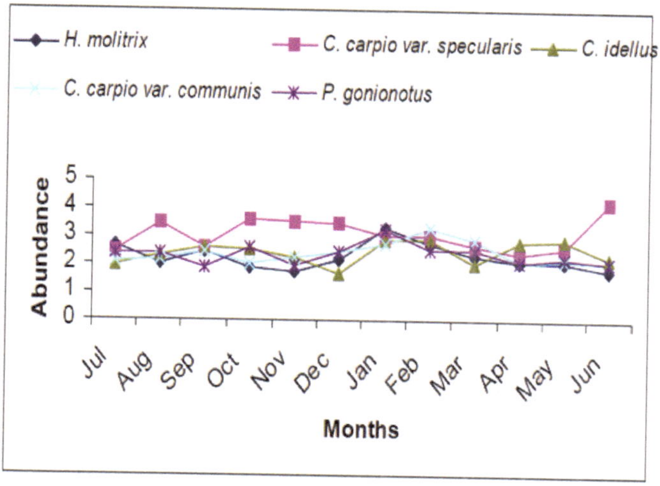

. Monthly average abundance of parasites in five exotic carp species (July 2004 – June 2006)

IMPACT OF PARASITES ON FISHES

The impact of parasites on fish may cause a significant host growth and development slow down resistance, and even cause death. The danger of parasitic disease mainly manifested in the following four aspects:

1. **Mechanical stimulation and injury:** Mechanical stimulation and causes tissue injury is a feature common parasitic disease can directly cause the death of fish or other lesions arising there from, adverse consequences.
2. **Extrusion and obstruction:** Some parasites in fish parasites can often result in the host tissues and organs of the squeeze, causing atrophy, necrosis and loss of physiological function.
3. **Predatory host nutrition:** Parasite nutrition from the host, the result would be more or less on the host create a hazard light are manifested as malnutrition, growth and development are affected, in serious cases can be death.
4. **The role of toxins:** Parasites in the parasitic process, its metabolites excreted in the host body. Some parasites can also secrete specific toxic substances, a certain impact on the host.

Parasites Of Exotic Carps

Kabata (1985) reported that shallow ponds and stagnant water favours the multiplicity of ciliate like Trichodina. Diseases affect the normal health conditions and cause reduction of growth, abnormal metabolic activates and even death. Health of any population depends on the control of disease and maintenance of a healthy relationship between living creatures and their environment (Snieszko 1983). Parasite acts either as a pathogen or vector for diseases (Roberts *et al.* 1986) .

Parasites Of Silver Carp

Helminths: *Bothriocephalus archeilognathi ,Dactylogyrus sp*

Protozoan: *Myxobolus pavlovskii , Trichodina sp.*

Crustacean: *Lernaea sp*

Parasites Of Common Carp

Helminth : *Dactylogyrus sp. , Gyrodactylus sp., Diplostomum sp., Phosthodiplostomum sp., Sanguinicola sp., Ligula intestinalis, Bothriocephalus acheilognathi, Contracaecum sp., Phylometra sp.*

Protozoa : *Ichthyobodo sp. , Ichthyophthirius multifiliis, Chilodonella sp., Trichodina sp., Eimeria sp, Myxobolus sp.*

***Crustacean* :** Ergasilus sp., Lernaea sp. Argulus sp.

Parasites Of Grass Carp

***Helminth* :** *Bothriocephalus sp., Dactylogyrus sp.*

***Protozoa* :** *Ichthyopthirius multifilis*

Copepod: *Sinergasilus sp.*

DIFFERENT PARASITES IN DETAILS

Bothriocephalus sp

Known world-wide from fish of the families Cyprinidae, Poecilidae, Cichlidae and Centrarchidae. The life history involves a definitive host, a fish and an intermediate host, a copepod. Heavily infected fish have a distended abdomen, (sometimes dropsy) Tapeworm-infected grass carp in China suffered from high mortalities (Yeh, 1955). Pathological changes, include lesions, inflammation of the intestine and severe "catarrhal-haemorrhagic enteritis". Infection can be readily detected from faecal material, revealing eggs, or residues of segments, and by autopsy.Infective stage is Procercoids .

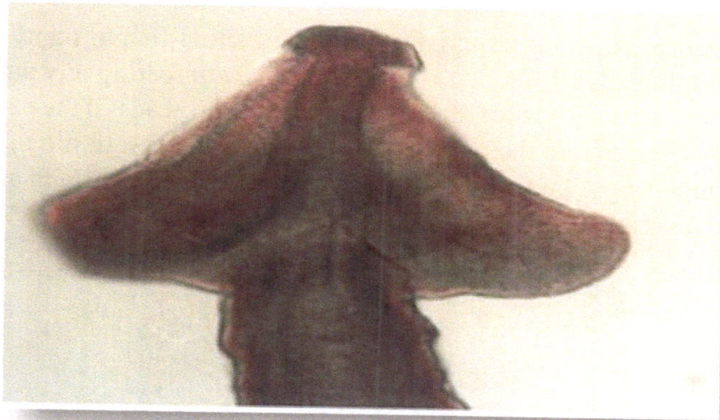

Fig 1 : *Bothriocephalus sp.*

Myxosporea sp

Very common in Cichlidae, Cyprinidae and in brackish waters in Mugilidae. Transmission involves tubificid oligochaetes as intermediate hosts. It Cause histozoic (*e.g.* in tissue) and coelozoic (in internal cavities *e.g.* in gall and urinary bladders) infections. Gross signs of histozoic infection whitish cysts with a milky substance containing microscopic (5-5 μm longer axis) spores. Small cysts present in viscera , connective tissue and muscles, Coelozoic myxosporea produce few spores, often only two. Most notorious is the whirling disease of trout, manifested by skeletal deformities. Myxobolus spp. caused locomotory disturbances coupled with emaciation, and sunken eyes in brain infections , anaemia and haemorrhagic dropsy and mortality in a heavy cardiac infection) and circulatory disfunctions in infections at the bases of the gill lamellae.

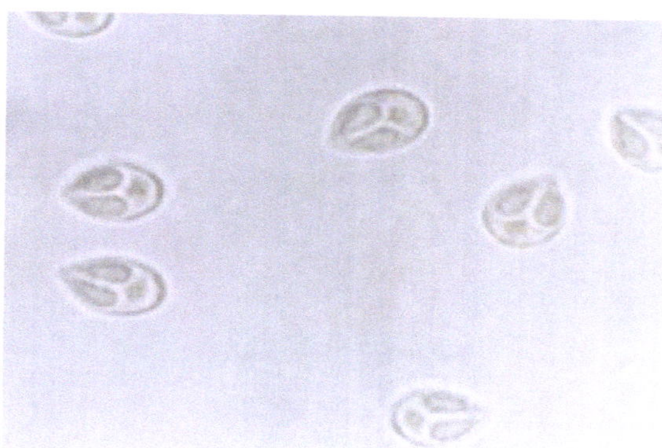

Fig 2 : *Myxosporea* sp.

Lernea sp

Lernaea cyprinacea, an opportunistic species infecting fish of many families as well as amphibians . Infecting gills, and females infecting any site on the fish integument, may harm the fish. In several cases the heads of the parasitized specimens penetrated into the body cavity and embedded itself into the liver in silver carp fingerlings (Jalali, 1987). Lernaeid species preferentially aggregate around the eyes, causing destruction of the lens and blindness .

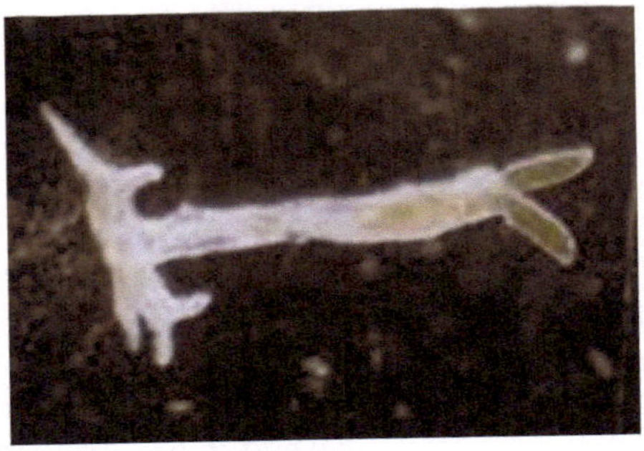

Fig 3 : *Lernea* sp.

Ergasilus sp.

In Ergasilidae only the female is parasitic, and is found on the gills of fish. Attached to gill filaments produce small foci of erosion. Excretion of proteolytic enzymes for external digestion. Irritation often results in responsive hyperplasia of the epithelium. Prolonged infection causes fusion and embedding of lamellae, resulting decrease in the respiratory function of the gills.

Fig 4 : *Ergasilus* sp.

Argulus sp

Argulids (fish lice). Argulids infecting the skin or mouth of fish induce severe local damage to the integument. Damage caused by piercing proboscis stylet which enters deep into the skin's dermal layer and secretes lytic and toxic substances resulting in acute haemorrhagic, inflamed wounds. Persistent irritation caused in heavy infections may affect fish appetite with resulting anorexia, and cessation of growth.

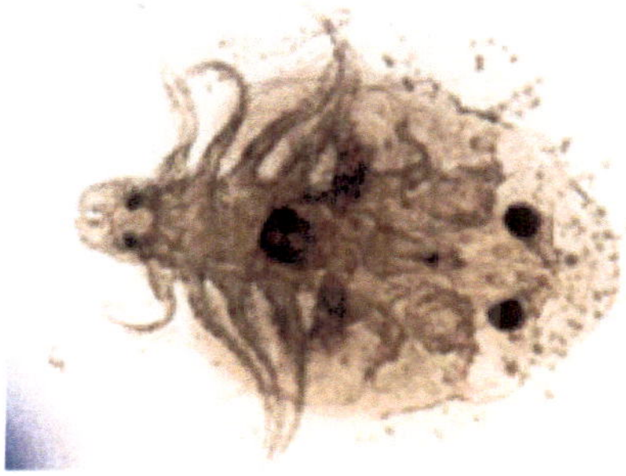

Fig 5 : *Argulus* sp.

Monogeneans

Gyrodactylidae are Viviparous .Dactylogyroidea are oviparous.Attachment organ is opisthator. *D. vastator* infection, in the gills of carp fry, induces severe hyperplasia of the gill filament . *D. lamellatus,* in grass carp, induces epithelial proliferation. Fish heavily infected with Gyrodactylus appear pale, due to excessive mucus secretion and epithelial proliferation. In the more heavily infected skin zones, there is skin erosion; desquamation of the skin epithelium, focal haemorrhagic lesions and loss of scales. Infection originated annually from "resting" eggs remaining in the ponds from the previous year's growing season.

Fig 6 : *Gyrodactylus* sp.

Fig 7 : *Dactylogyrus* sp.

Trichodina sp

Attached to the skin or gill of host with its adhesive disc. Trichodina can cause infection to fish of any developmental stages. Infestation in jeapodise juvenile fish causes severe mortality. Causes severe mucus secretion in fish, the more intensive parts such as fins, head, body surface and so appeared a layer of white shade, known as white-headed white-mouth disease.

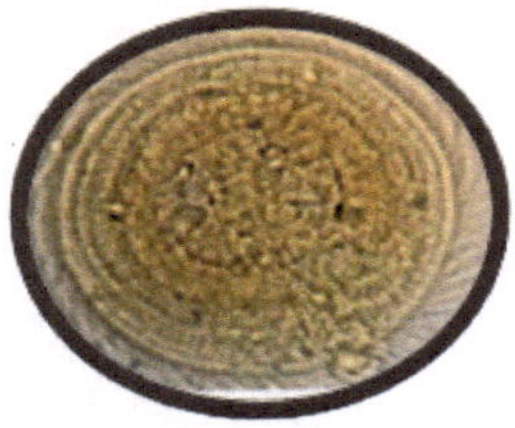

Fig 8 : *Trichodina* sp.

Other ciliates

Ichthyobodo necator attaches itself to epithelial cells and through an inserted protrusion consumes their contents. Mortalities occur in fish overcrowded in holding tanks, ponds and in both warm and cold water conditions. *Chilodonella sp.* and *I. necator,* in overwintering carp, but occurred more often in relation to other stress factors such as high levels of overcrowding or high nitrite concentrations. Cellular destruction primarily occurs due to direct action of the parasites, and later by enhanced abrasion of the peripheral cells after the depletion of mucus forming cells. The production of mucus cells is limited and the infected fish become "dry". Some parasites seem to yield cytotoxins or proteolytic enzymes which could be the cause of spongiosis in fishes.

Ichthyopthirius sp.

Most species of freshwater fish are susceptible. Infective stage is theront. Gross signs - white spots on the skin and the gills. Chronic infection will cause serious damage to the skin, fin and gills; corneal infection impairs vision. In some infections, parasites cause widespread lysis of the inner layer of the epithelium. Spontaneous recovery has been observed in both natural infections in natural habitats and in aquaria.

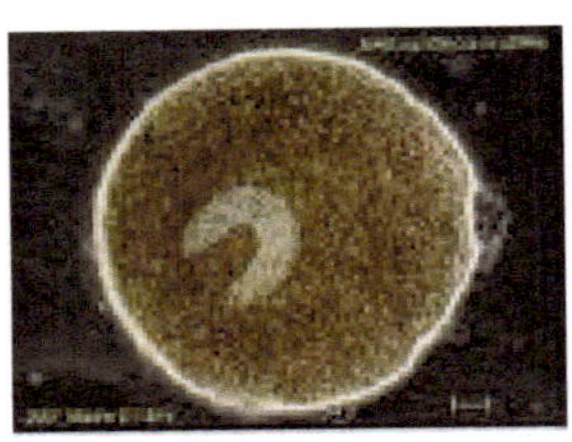

Fig 9 : *Ichthyopthirias* sp.

Digenean

Trematodes or Digenea are flatworms (Platyhelminthes), with a multiple host life cycle and require a mollusc as their first Intermediate host. Suckers are used for attachment and locomotion. Most trematodes are hermaphrodite. Cercariae of these flukes actively penetrate into their definitive piscine host. Adult trematodes, infecting the digestive tract of fish, are considered harmless, even when their numbers are high.Extraintestinal trematode infections, are potentially pathogenic. Adult worms and trapped eggs can physically obstruct the passage of blood, causing thrombosis and subsequent necrosis. Escape of miracidia through the gill epithelium causes anaemia.

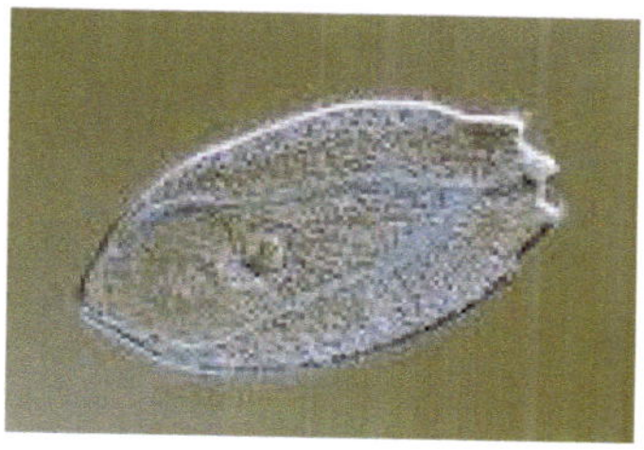

Fig 10 : *Diplostomum* sp.

Nematodes

Most economically important helminth parasites of fishes in the world. Nematoda (round worms) are very distinctive in shape. Fish acts as definitive intermediate host or paratenic hosts for nematodes .Most adult forms are large enough to be visible to the naked eye. The majority occur in the alimentary system and only a few enter tissues or inner cavities. Aged and dying worms locked in the abdomen or trapped in tissue provoke severe inflammatory responses, granuloma and fibrosis .

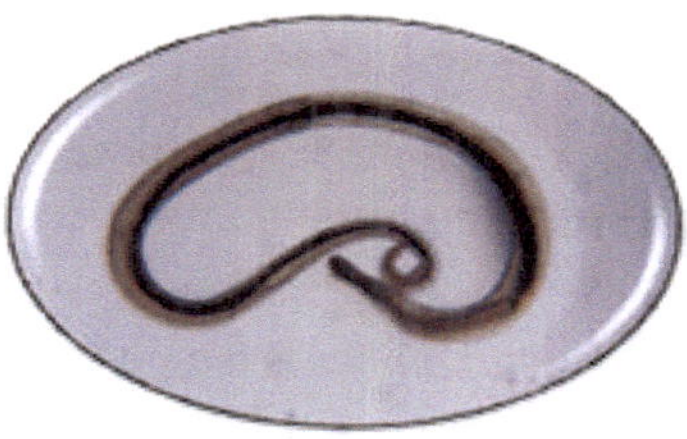

Fig 11 : *Camallanus* sp.

FACTORS AFFECTING PARASITISM

- Biotic Factors:
- Symbiosis

- Competition
- Predator- Prey Relationship
- Abiotic Factors
- Temperature
- Season
- Crowding Etc.,

BIOTIC FACTORS:

Symbiosis – A relationship in which two organisms of different species 'live together' for a period of time

Competition – A relationship where two types of organisms compete for the same resource such as food, water, nesting

Predator / Prey Relationships – The relationship in which one organism (predator hunts and eats another (prey).

ABIOTIC FACTORS:

Temperature

The Asian tape worm (bothriocephalus) seems to be a thermophilic species (Hoffman, 1980). Low temperatures delay or even interrupt development and the life cycle. Davidov (1978) found that temperatures below 15°C will delay development to 6–8 months.

Larval development of *L. cyprinacea,* is temperature dependent. Temperatures of 30°C seem to be the upper limit for optimal development. only three or less generations are produced, in winter as compared to seven or eight generations in summer.

Eggs of Dactylogyrus species from carp hatch within two to six days at temperatures of 20–28°C.

Higher temperature hasten trophont maturation and tomont division in Ichthyopthirias. In lower ambient temperatures the survival of the tomites is prolonged in Ichthyopthirias.

Egg development of the *A. japonicus* is temperature dependent. Embryonic development is interrupted at 10°C and larval development is completely interrupted below 12°C .

Host Size

Fish fingerlings become more susceptible to pathogen because of their immature immune system (Anderson, 1974). Nie and Kennedy (1991) reported higher infection levels in larger fish because it provides more space.

Infection of bothriocephalus related not only to worm burden, but also to the size and condition of the host. Although the Asian tape worm is not fastidious in the choice of host. It may have different degrees of compatibility with different host species, which may also be expressed as variation in tolerance and defence response among infected fish species. Copepodite infections are potentially lethal to small fish, while large fish (such as Bagrus) will tolerate extremely heavy gill infections . The adult parasites of lernea are particularly harmful to large fish because of their large size, mode of attachment and feeding. Gyrodactylids, are, pathogenic to younger fish and in intensive culture conditions.

Stress

Stress is an abnormal physiological condition of fish resulting when its collective adaptive responses to an environmental factor(s) are extended to or approaching the fish's limit of tolerance for that factor. Health conditions of fish populations are direct products of management. The majority of fish health problems in fish populations, especially disease problems, are directly linked to stress on the fish from some environmental factor(s). Direct relationships exist between environmental quality, fish health and disease. Avoiding stress by maintaining good environmental quality through proper management is essential to the maintenance of a healthy, disease free fish population. Thus, emphasis in order of importance is on stress (from environmental quality factors), health and disease. Obviously, not all environmental stressors are under management control, and some obligate infectious diseases are not predisposed by an environmental stressor. Nevertheless, these and all health problems are subject to some measure of management.

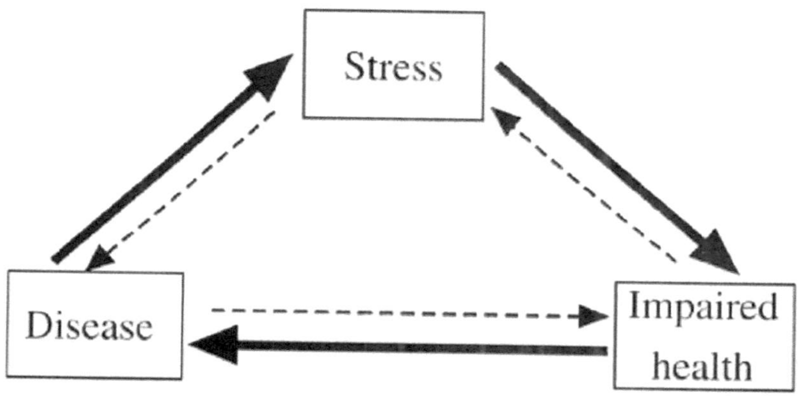

Fig 12 : Illustration of the direct interrelationships between fish stress, health and disease.

Crowding

High stocking density induces bio-ecological stress to fry and make the fry more susceptible to the parasitic infestation (Sniesko, 1974). Crowding fish into culture systems or aquaria often promotes infestation of monogeneans.

Feeding Habits

Feeding habits of the host account for a large percentage of the variation in the total number of parasites per host species (Williams & Jones 1994). The number of parasites depend on how much the host eats and whether the food is infected. As soon as the breeding season ends fish start eating which results in an increase in abundance of parasites from autumn to winter.

Seasonal Changes

Ahmed et al. (1991) observed that the prevalence of diseases was more in the winter season every year. As the water quality parameters fluctuate very quickly during winter and summer season, fish becomes affected with diseases in these two seasons. Infections of bothriocephalus assume a clear seasonal pattern, with peak incidence in the summer. Infestation seems to become interrupted during the coldest part of the year as is evident from the complete absence of young scolices.

Dissolved Oxygen

Egg development of lernea seems to be affected as levels of dissolved oxygen decline. *D. ranarum,* possessing haemoglobin, seems to be capable of withstanding low levels of dissolved oxygen.

Breeding Season

During the breeding season of fish lesser number of parasites invade the host because of the presence of the oestrogen (Rahman and Jahan (2002).

Haematological Changes

During the Myxobolus infection, damage to the RBCs and leukocytes has also been detected. Protozoan infection and subsequent infection with microorganisms seem to be responsible for the damage to the RBCs and WBCs. Decrease in haematocrit values occur in case of cryptobia infection. In case of bothriocephalosis the hemoglobin content get reduced . in case of Helminth infection the RBCs and haemoglobin content of the blood get reduced and the WBCs get increased. The WBC get increased to protect the fish from the parasitic infection. The parasites feed on the blood which results in the reduction of haemoglobin.

Histological Changes

GILLS : Dactylogyrus infestation causes degenaration of the gills, holes in the piercing areas, telangiectasia of the capillary of the lamellae.

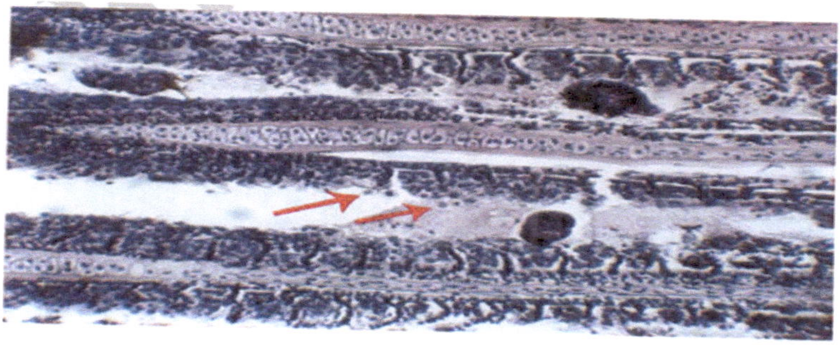

Fig 13 : Degeneration in the gills of common carp due to Dactylogyrus infection

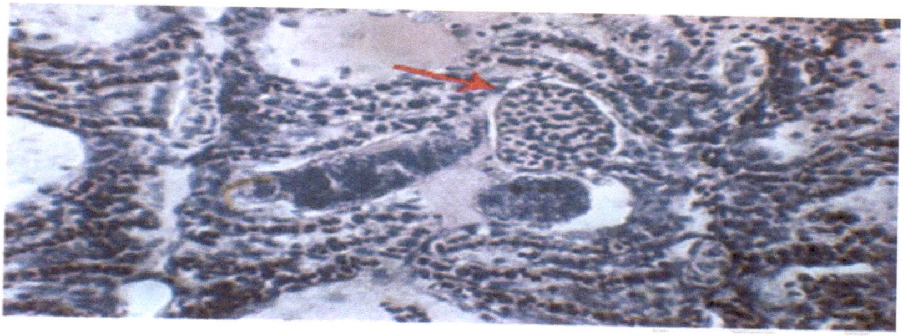

Fig 14 : Telangiectasia of the capillary of the lame

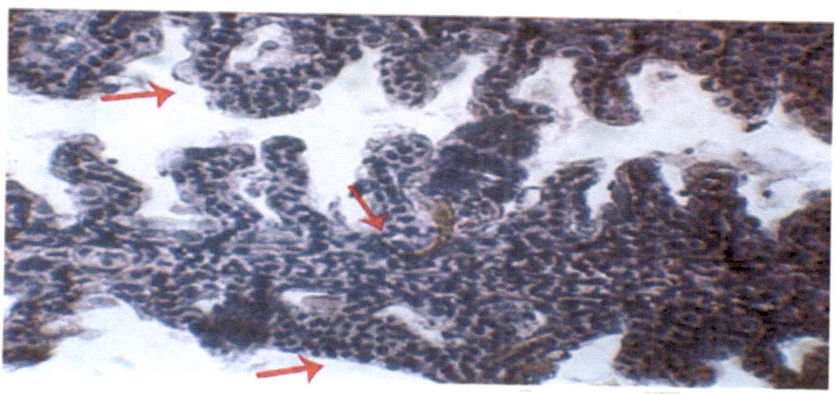

Fig 15 : At the attachment site of the *D. sahuensis* deep holes are formed due to degeneration of the lamellae 3 to 5 lamellae grows together.

Kidney : Kidney sections taken from protozoan infected common carp shows lymphoid cells prevailed In glomerular capillaries and an increase of interstitial lymphoid tissue was observed when infected carp were compared to uninfected carp.

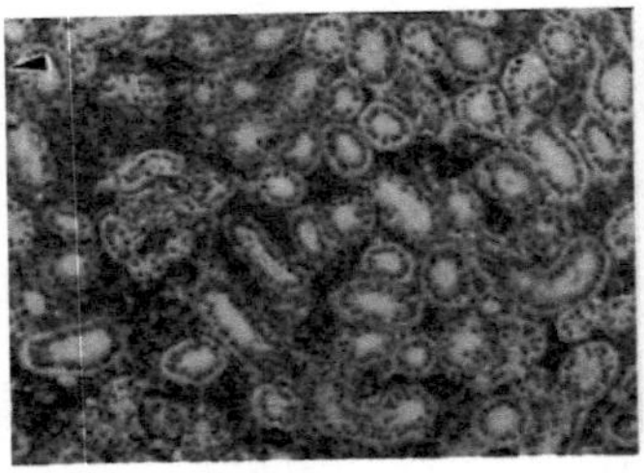

Fig16 : Uninfected kidney

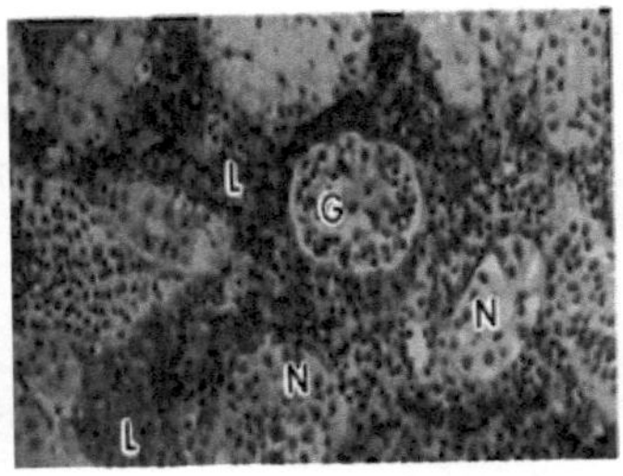

Fig 17 : Infected kidney

Scales

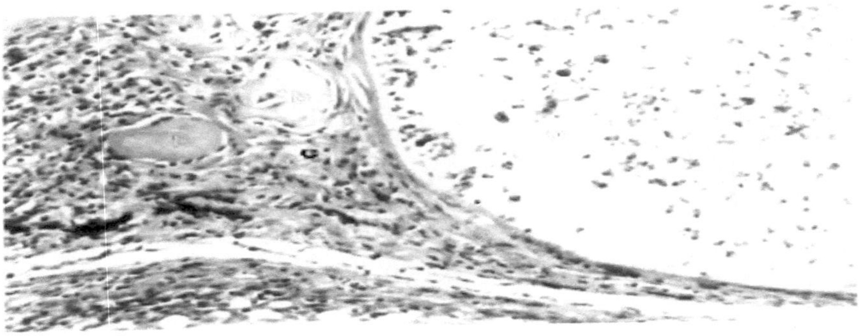

Fig 18 : Myxosporean infection causes deformation of the scales in common carp.

10

Parasites Picture Gallery

An extraordinary variety of viruses, bacteria, and organisms stand ready to attack us and feed off our bodies' cells. Using microscopes of varying power, photographers show us a teeming microbial world that we could barely imagine without their help.

MEASUREMENT OF MICROBES

The approximate sizes of microbes can be measured by using the following rule of thumb:

- **Viruses** are the smallest of all infectious agents, averaging about 100 nanometers (100 billionths of a meter) in length. They have so few genes and proteins of their own that in order to reproduce they need to commandeer the machinery of the cells they invade.
- **Bacteria** vary widely in size and shape, but tend to be at least 10 times larger than viruses, or at least 1 micrometer (1 millionth of a meter) long. They are single-cell organisms that reproduce independently.
- **Single-cell Organisms** tend to be at least 10 times larger than bacteria, or about .01 millimeter long.
- **Multi Cellular Organisms** are so large they can usually be seen with the naked eye. Tapeworms, for instance, can reach a length of 6 meters (20 feet).

Food and water are the most common sources of invading organism transmission. Since most of us eat three times a day and drink water frequently throughout the day, our exposure to these sources is constant. Tap water has been found to be contaminated with harmful organisms. Both plant and animal foods carry organisms, and cleaning and cooking methods often do not often destroy them before ingestion. The CDC (Center for Disease Control) cites food as the catalyst behind 80 percent of the pathogenic outbreaks in the U.S. Most are linked to restaurants and delis where less than sanitary conditions exist — from food preparation and storage to the utensils and servers' hands.

Animals, just like humans, can become infected with unhealthy organisms. Internally, contaminated water and food can spread the problem to our pets. Externally, animals become infected by organisms on their bodies, especially on their fur, because of exposure to infected animal wastes. Forgetting to wash your hands even one time after handling or cleaning up after your animal can transmit the parasite to you. Pets are a wonderful part of our lives. They provide comfort, companionship, protection, amusement, and unconditional love for their owners. Yet, pets, like humans, are often victims of serious infections that can unintentionally be passed on to their owners. In fact, there is a whole set of diseases classified as 'zoonoses' (animal-transmitted diseases) in parasitology textbooks. Animals are major carriers of harmful organisms, and most physicians, let alone the general public, are seemingly unaware of this fact. Experts have projected that of the 110 million pet dogs and cats in this country, over half may be infected with at least one or more different kinds of harmful organisms. Considering these numbers, the potential for transmission of invading organism infection from animals to humans is extremely high.

The CDC estimates that the number of parasites present in the United States alone number in the thousands. These harmful organisms are biochemically complex creatures in their life histories, development, reproductive cycles, nutritional requirements, and manifestation. They are categorized according to structure, shape, function, and reproductive ability. These include microscopic organisms (protozoa); roundworms, pinworms, and hookworms (nematoda); tapeworms (cestoda); and flukes (trematoda).

PROTOZOA

Making up approximately 70 percent of all invading organisms, protozoa are invisible to the naked eye. They are one-celled microscopic organisms, but don't let their size fool you. Certain protozoans, through their intensely rapid reproductive ability, can take over the intestinal tract of their host; and from there go on to other organs and tissues. Some feed on red blood cells. Some protozoa produce cysts - closed sacs in which they may be safely transported through food and water from one person to another. In the cyst state, protozoans are safe from destruction by human digestive juices. These one-celled 'vampires' can actually destroy the tissues of their hosts. According to experts, an estimated 7 million people across the U.S. have some form of protozoa living inside of them. Common protozoans include: *Endolimax nana*, *Giardia lamblia*, *Entamoeba histolytica*, *Cryptosporidum parvum*, *Blastocystis hominis*, *Trichomonas vaginalis*, *Toxoplasma gondii*, *Cyclospora cayetanensis*, *Cryptosporidium muris*, *Pneumocystis carinii*, *Plasmodium malariae*, *Plasmodium ovale*, *Plasmodium vivax*, *Plasmodium falciparum*, *Leishmania donovani*, *Leishmania tropica*, and *Leishmania braziliensis*.

NEMATODE

While the protozoans are only single-celled, nematode creatures are multi cellular. The adult worms multiply by producing eggs called ova or larvae. The eggs usually become infectious in soil or in an intermediate host before humans are infected. It is interesting to note that unless the worm infection is heavy, many individuals do not show signs of infestation. While it may be unpleasant to consider, it is true that the human host can coexist quite comfortably with a few worms, unless they reproduce in great numbers and create organ obstruction. Experts claim that 'some type of worm is already in the intestines of over 75 percent of the world's population'. This is a frightening statement. Common nematodes include: *Roundworm (Ascaris lumbricoides), Hookworm (Necator Americanus, Ancylostoma duodenal)*, P*inworm (Enterobius vermicularis), Roundworm (Toxocara canis, Toxocara cati), Heart worm (Dirofilaria immitis), Strongyloides (Stronglyoides stercoralis), Trichinella (Trichinella spiralis), Filaria (Wuchereria bancrofti, Brugia malayi, Onchocerca volvulus, Loa loa, Mansonella streptocerca, Mansonella perstans, Mansonella ozzardi)*, and *Anisakine larvae.*

CESTODE

Among the oldest known invading organisms, tapeworms are considered humanity's largest intestinal inhabitant. They each have a scolex (head) that attaches to the intestinal wall. As long as the head remains attached to the intestinal mucosa, a new worm can grow from it. Tapeworms do not contain digestive tracts but get their nourishment by absorbing partially digested substances from the host. They are whitish in color, flat, and ribbon-like, with a covering that is a transparent skin-like layer. Common cestoda include: *Beef tapeworm (Taenia saginata), Pork tapeworm (Taenia solium), Fish tapeworm (Diphyllobothrium latum)*, and *Dog tapeworm (Dipylidium caninum).*

TREMATODE

Trematode are leaf-shaped flatworms also known as flukes. They are parasitic during nearly all of their life-cycle forms. The cycle begins when larvae are released into freshwater by infected snails. The free-swimming larvae can then directly penetrate the skin of the human host or are ingested after encysting in or on various edible, vegetation, fish, or crustaceans. Common trematode include: *Intestinal fluke (Fasciolopsis buski), Blood fluke (Schistosoma japonicum, Schistosoma mansoni) Schistosoma haematobium), Liver fluke (Clonorchis sinensis), Oriental lung fluke (Paragonimus westermani)*, and *Sheep liver fluke (Fasciola hepatica).*

PICTORIAL GALLERY

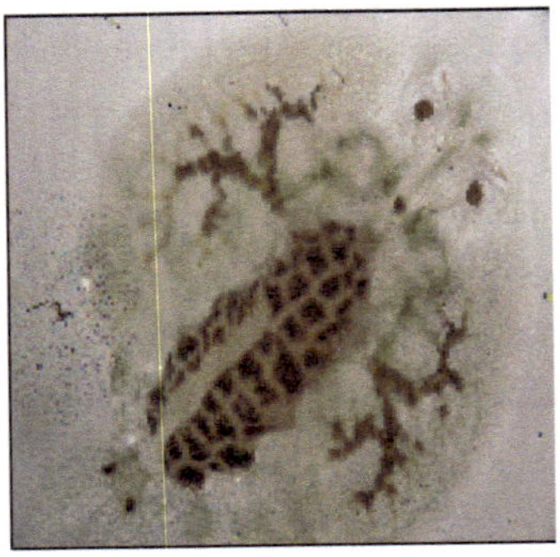

Argulus

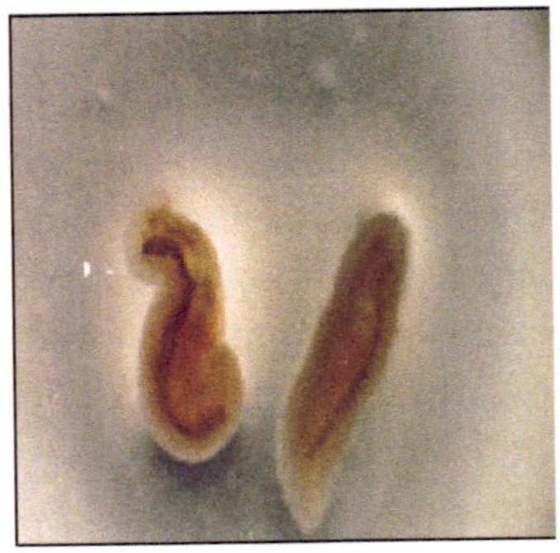

Digenean larvae

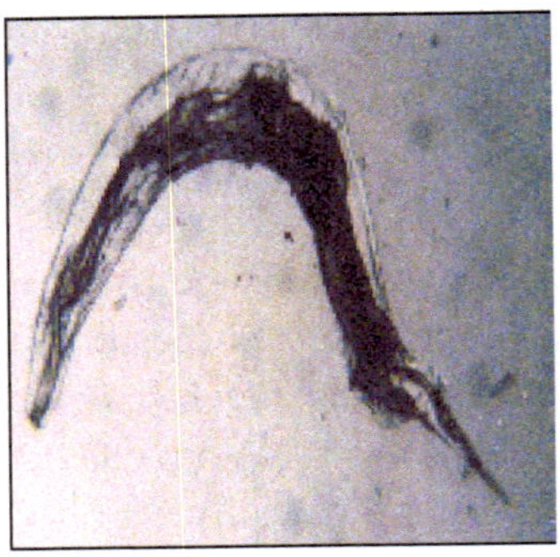

Indocucullanus sp.

Rhabdochona sp.

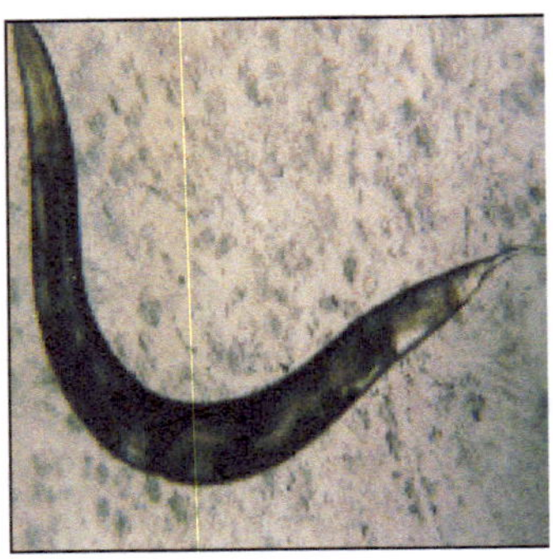

Indocucullanus sp.

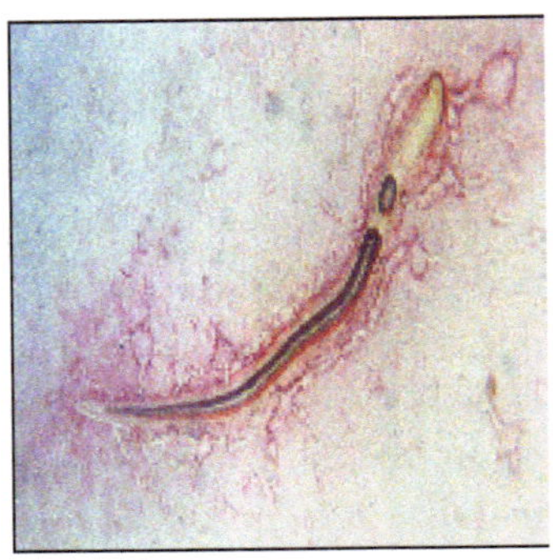

Camallanus sp.

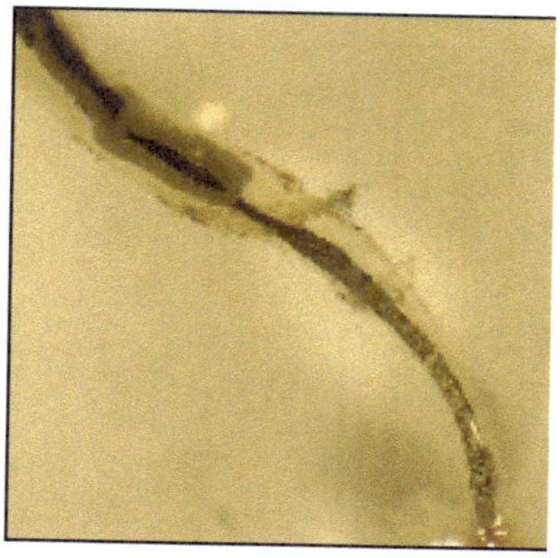

Lemea in Guppy

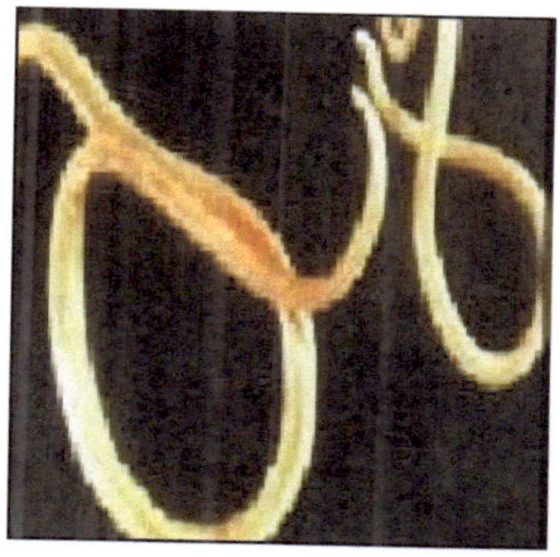

Round worm

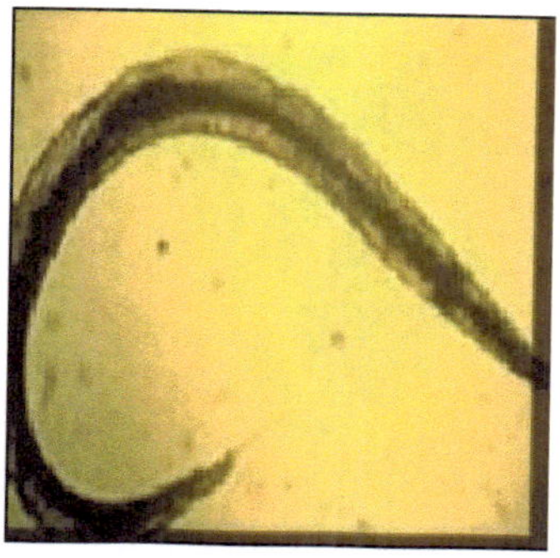

Pin worm

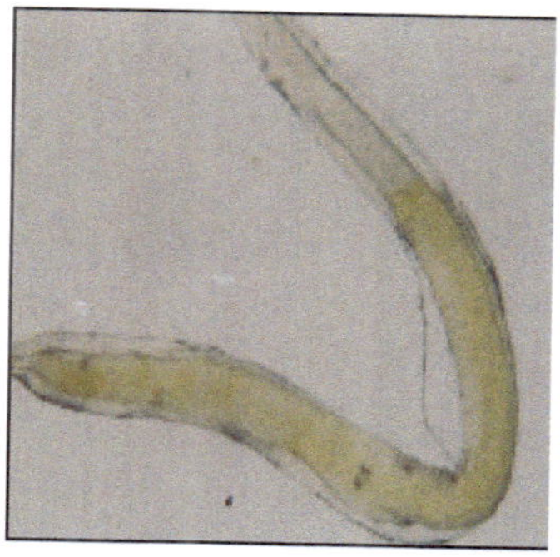

Wetmount of nematode
(Camallanus sp.*)*

Wetmount of nematode
(Camallanus sp.*)*

Camallanus sp.

Wetmount of nematodes *(Camallanus* sp.*)*

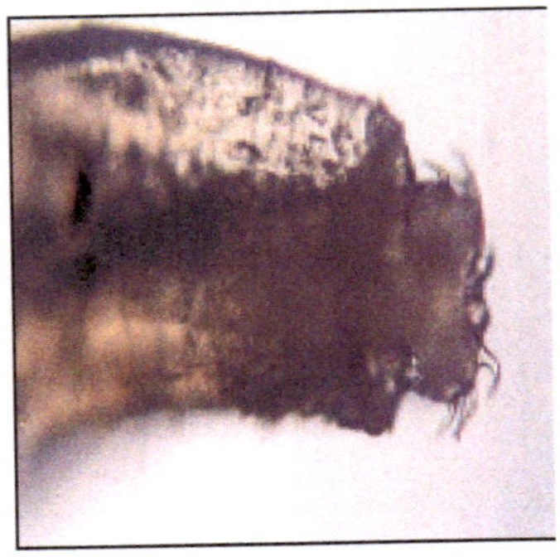

Wetmount of acanthocephala *(Pallisentis* sp.*)*

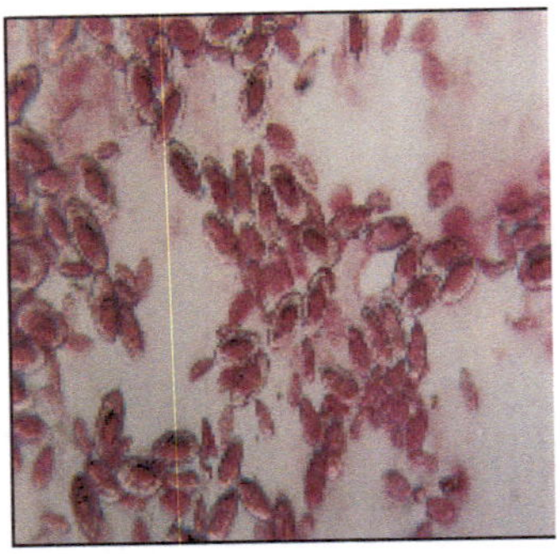

Wetmount of acanthocephalans eggs from intestine of Labeo rohita

(Nematode) Philanoma sps

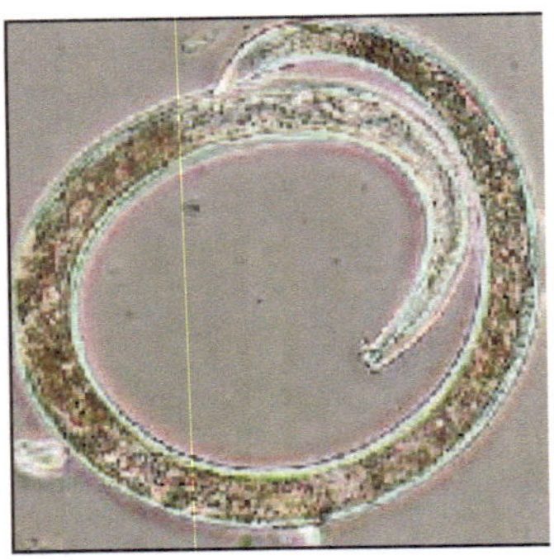

Camallanus sp.

Adult Bchinorhynchus salmonis

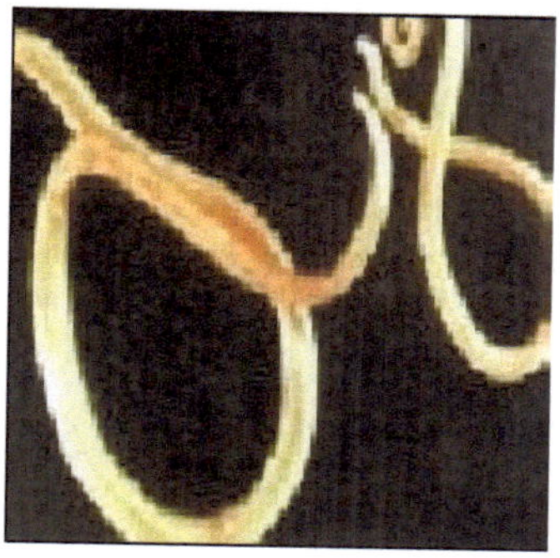

Roundworm

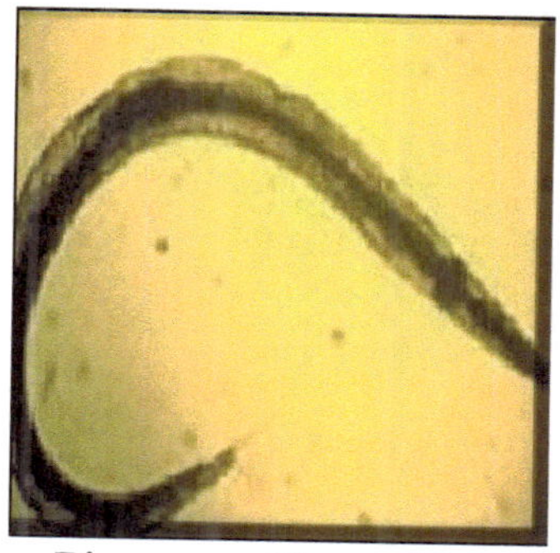

Pinworm *(Enterobius vermicularis)*, Zoonotic Parasite

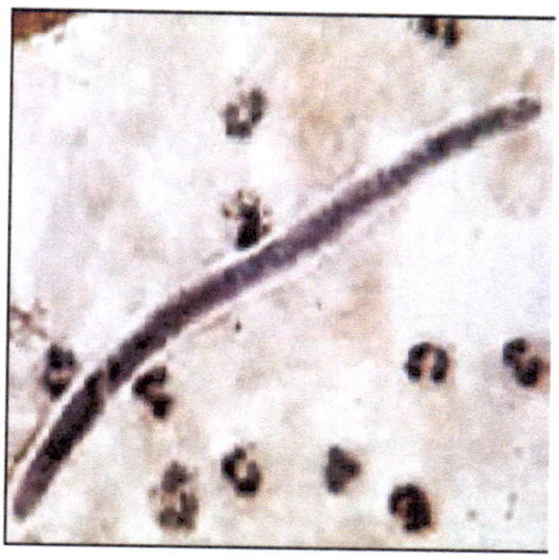

Trichinosis *(Trichinella spiralis)*

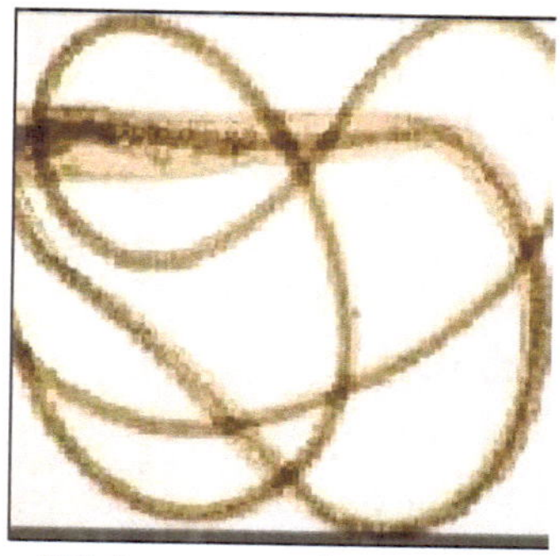

Whipworms *(Trichuris trichiura* and *Trichuris vulpts*

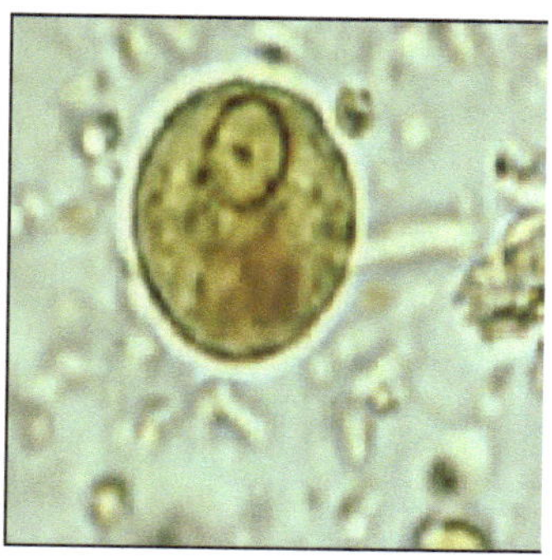

Protozoan Infections: *(Entamoeba hisolytica)* Amebiasis

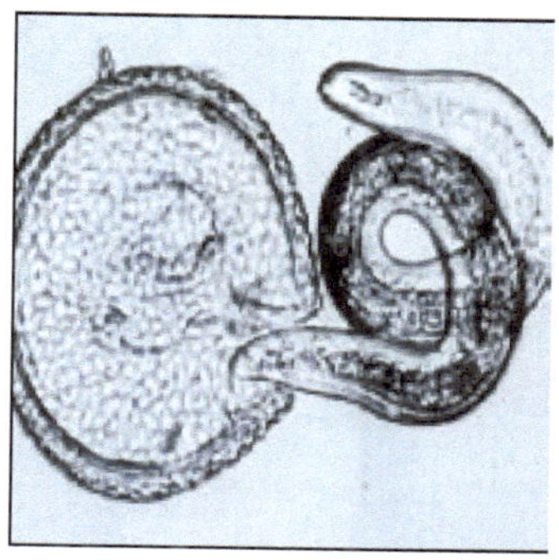

Toxocariasis *(Toxocara canis* and *Toxocara cati)*

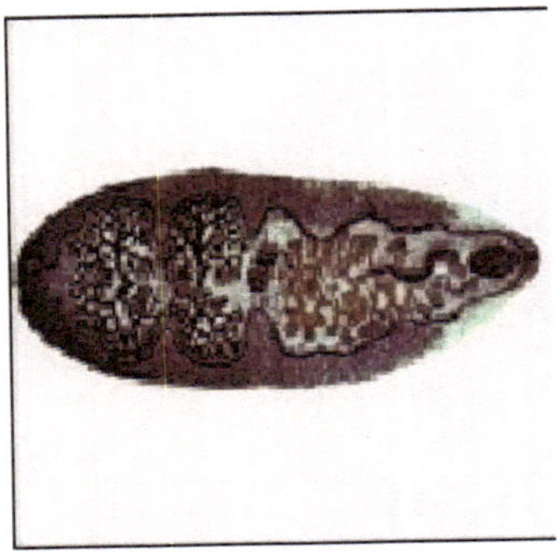

Intestinal Fluke

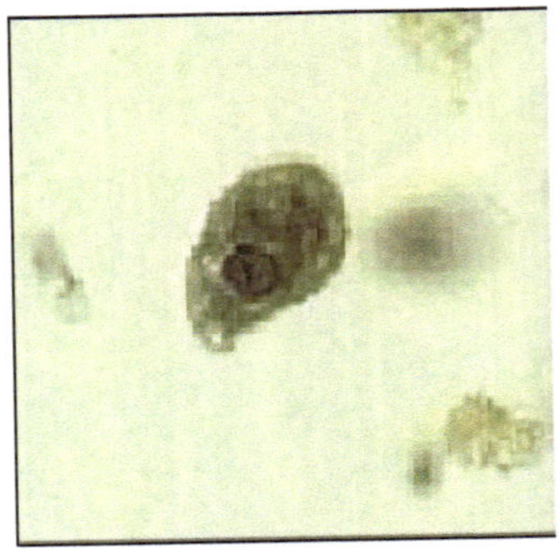

Amoeba Organisms

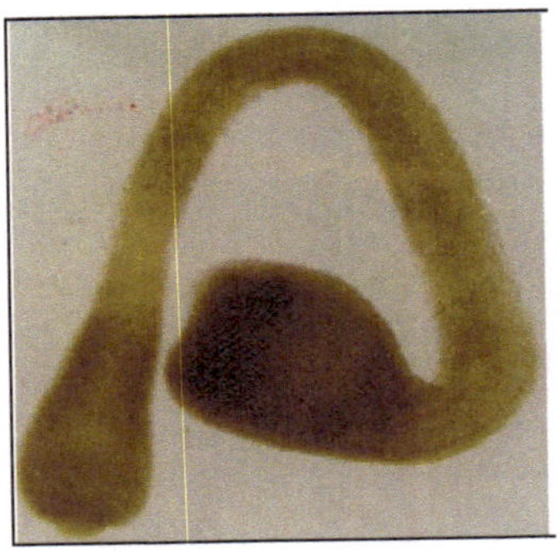

Wetmount of cestode
(Caryophyllaeus sp.)

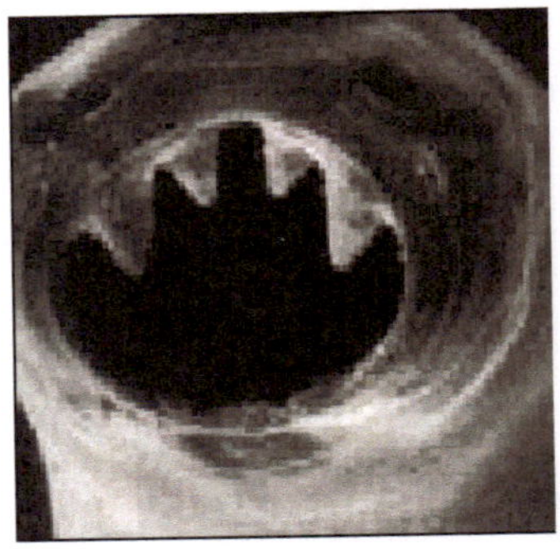

Hookworm

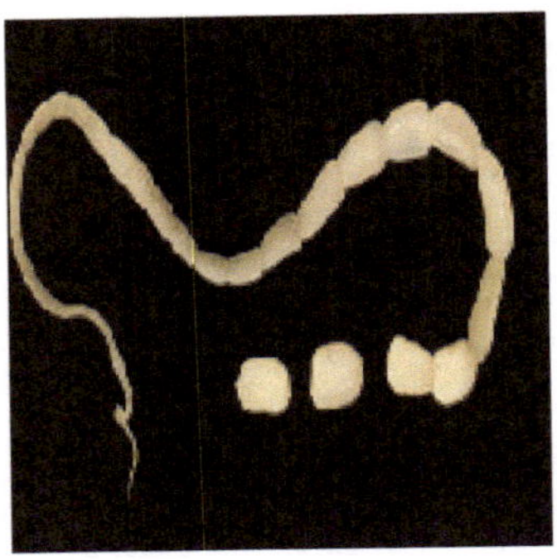

Tapeworms
(Dipylidium caninum)

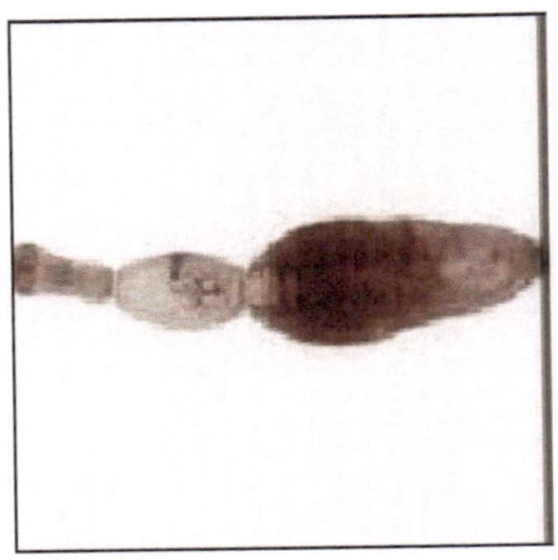

Dwarf Tapeworm

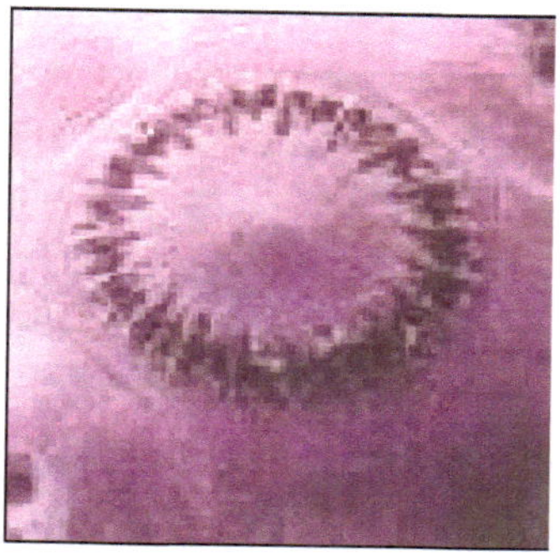

Fish tapeworm

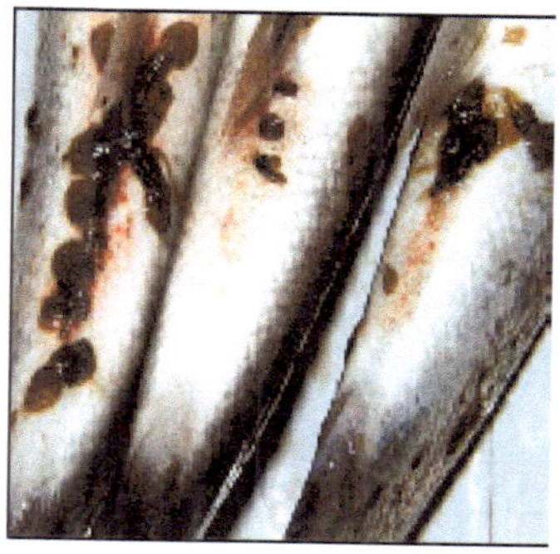

Fish lice

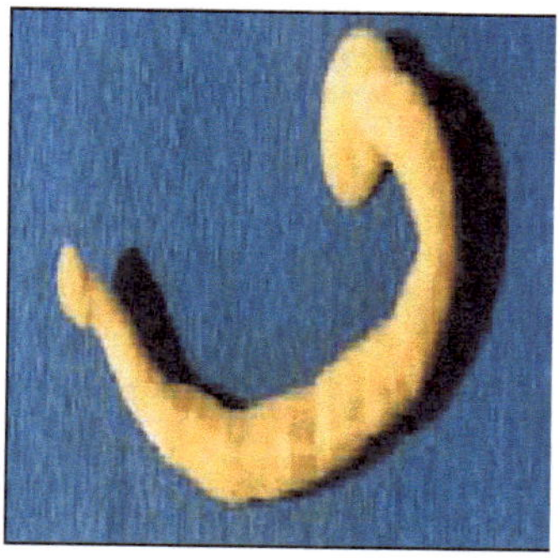

Leech *(Piscicola milneri)*

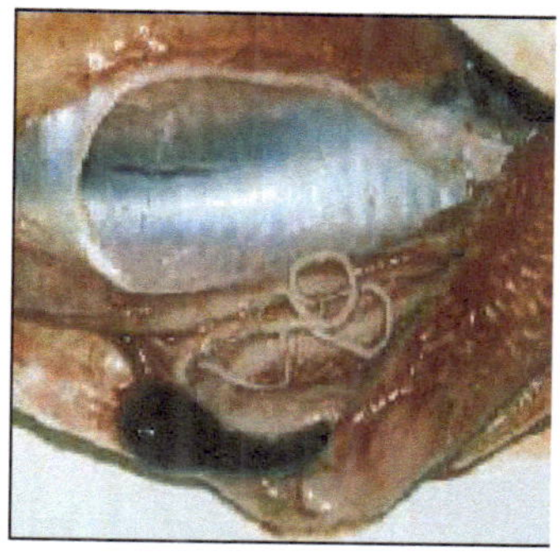

Nematode
Philanoma sps

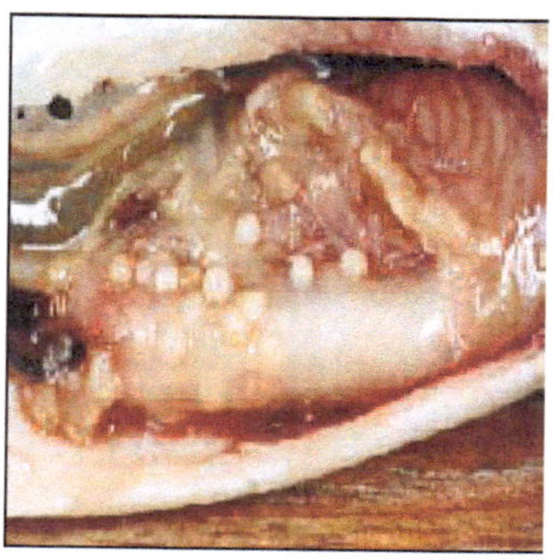

Cysts of
Diphyllobothrium sps.

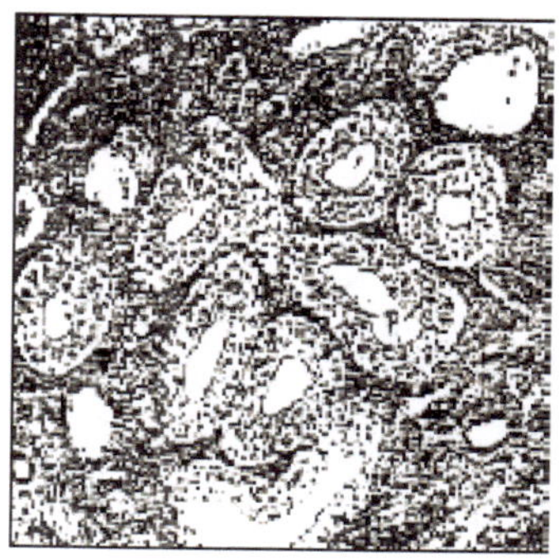

Microspores
Developmental stages

Caligoid copepods

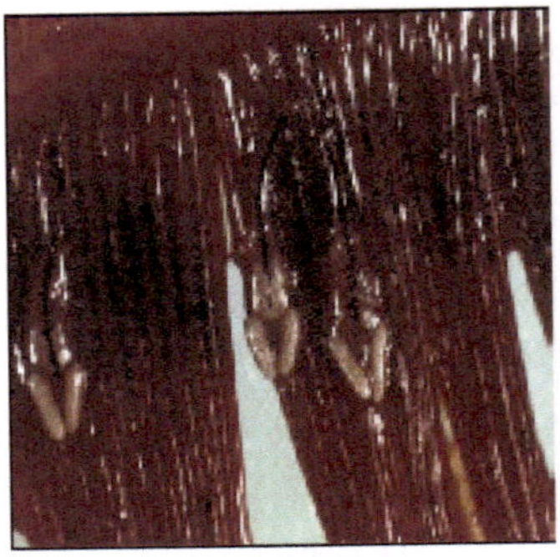

Copepods on fish gill

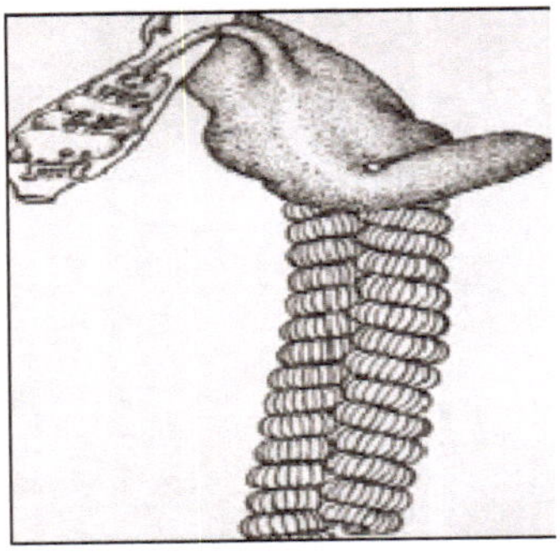

Copepod in Fish Heart

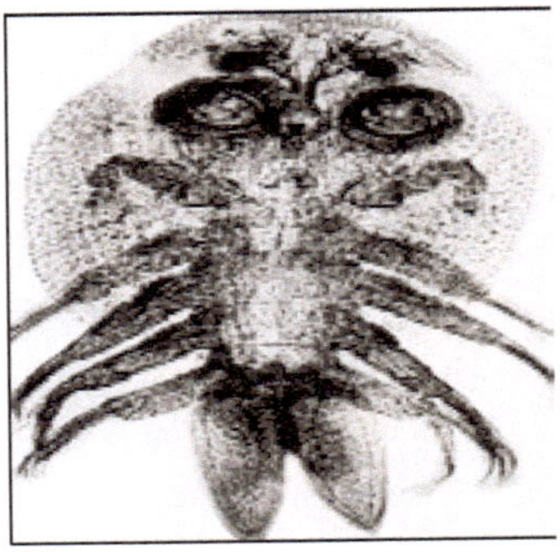

Fish Louse
Magnified view

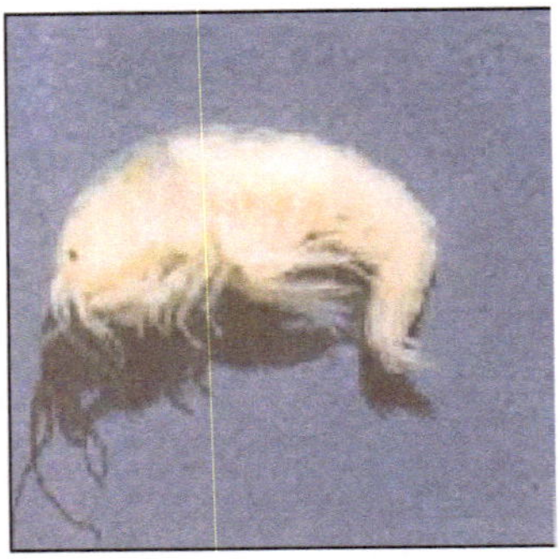

Acanthocephala (Spiny headed worms)

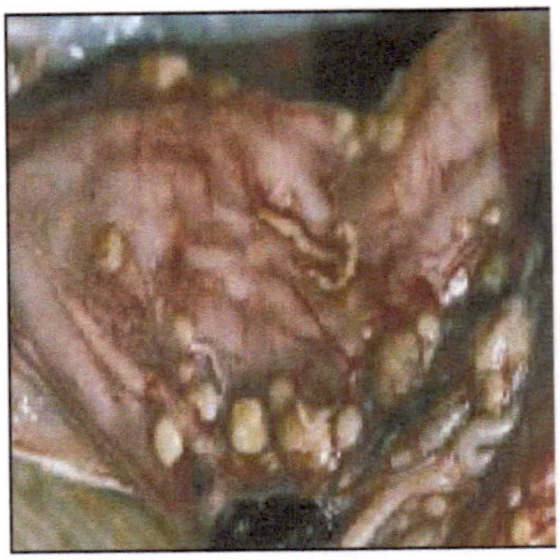

Diphyllobothrium larvae encysted and free

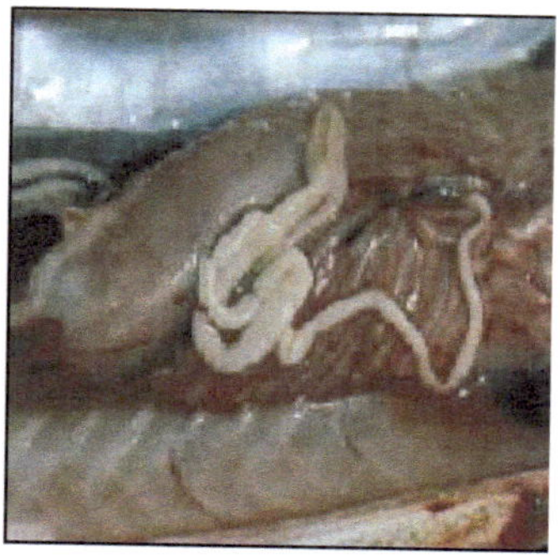

Adult tapeworm,
Eubothrium sp

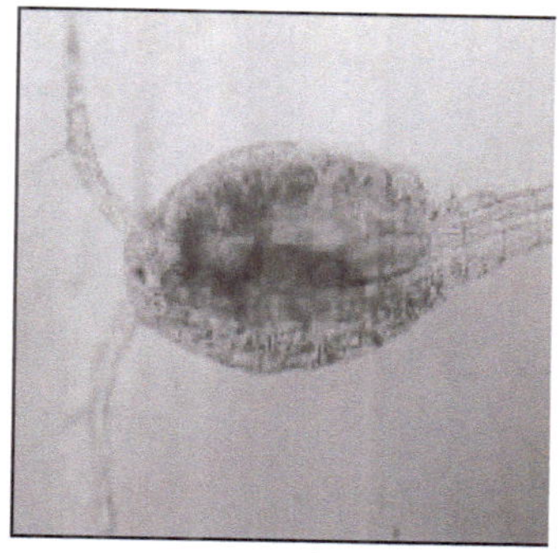

Copepod infected with
larval procercoids

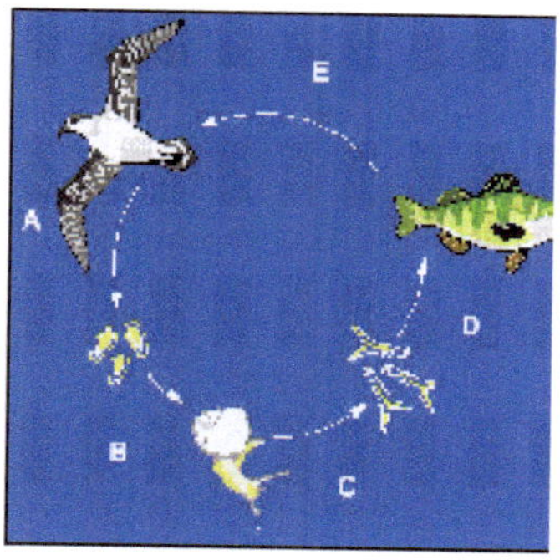

The Life Cycle
(*Diplostomum* sp.)

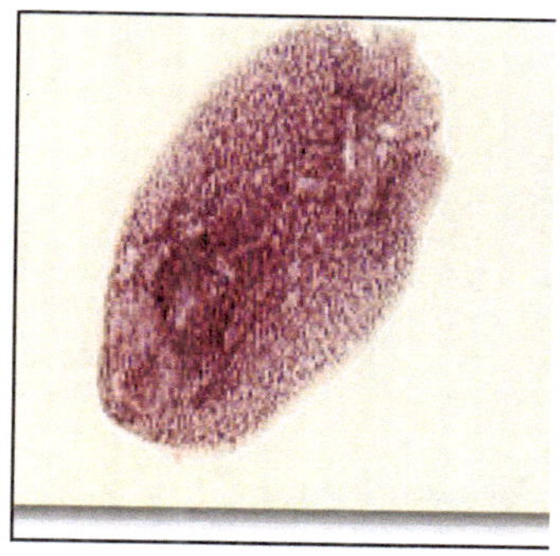

Diplostomum sp.

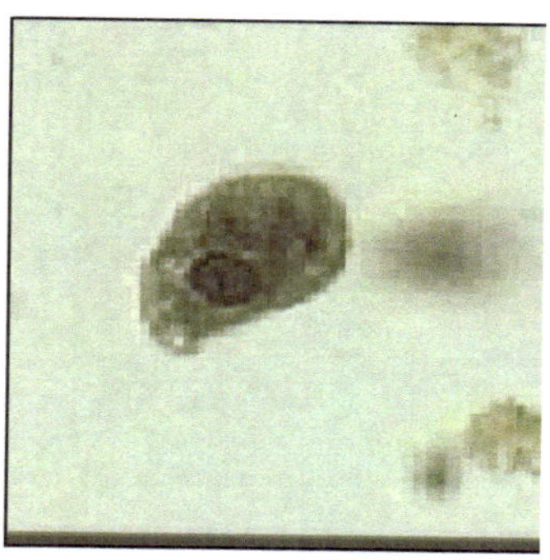

Amoeba sp.

Costia sp. in fish

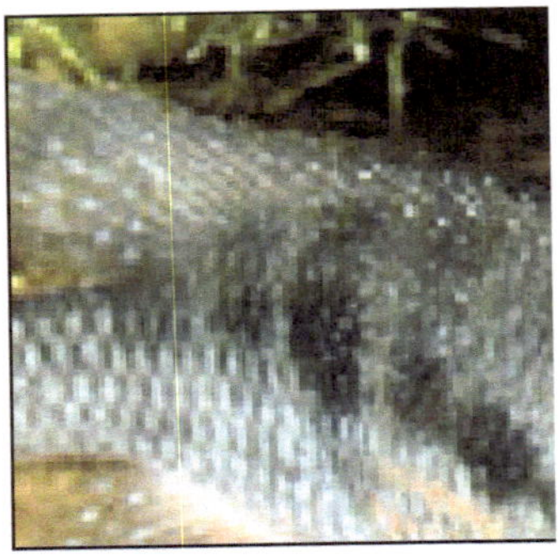

Ich *(Ichthyophthirius multifilliis)* in fish

Fish Lice

Hexamita (Hole-in-the-Head)

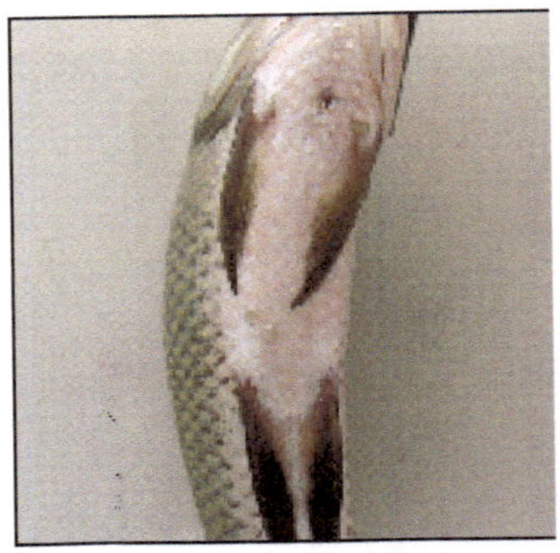

Argulosis in Lobeo rohita

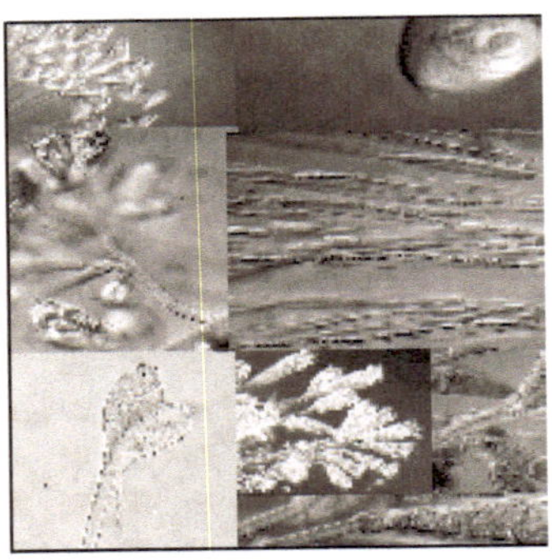

Epistylis sp.

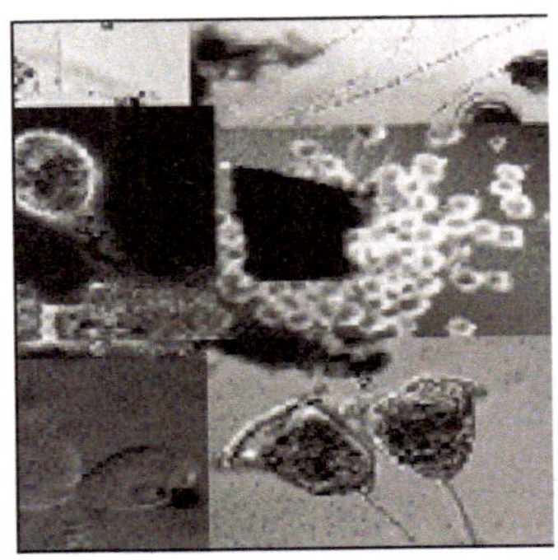

Vorticella sp.

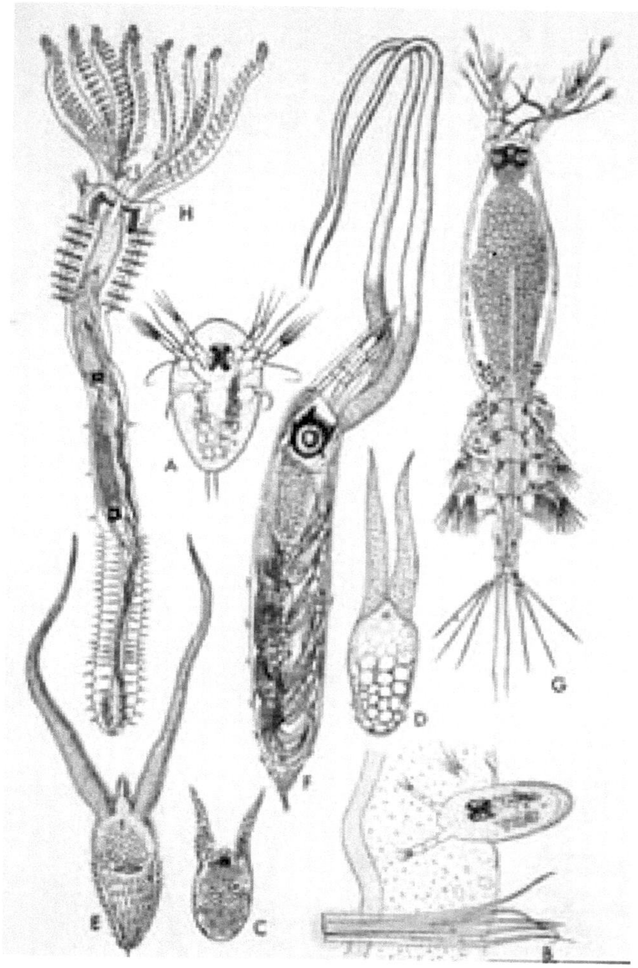

Polychaete Worms

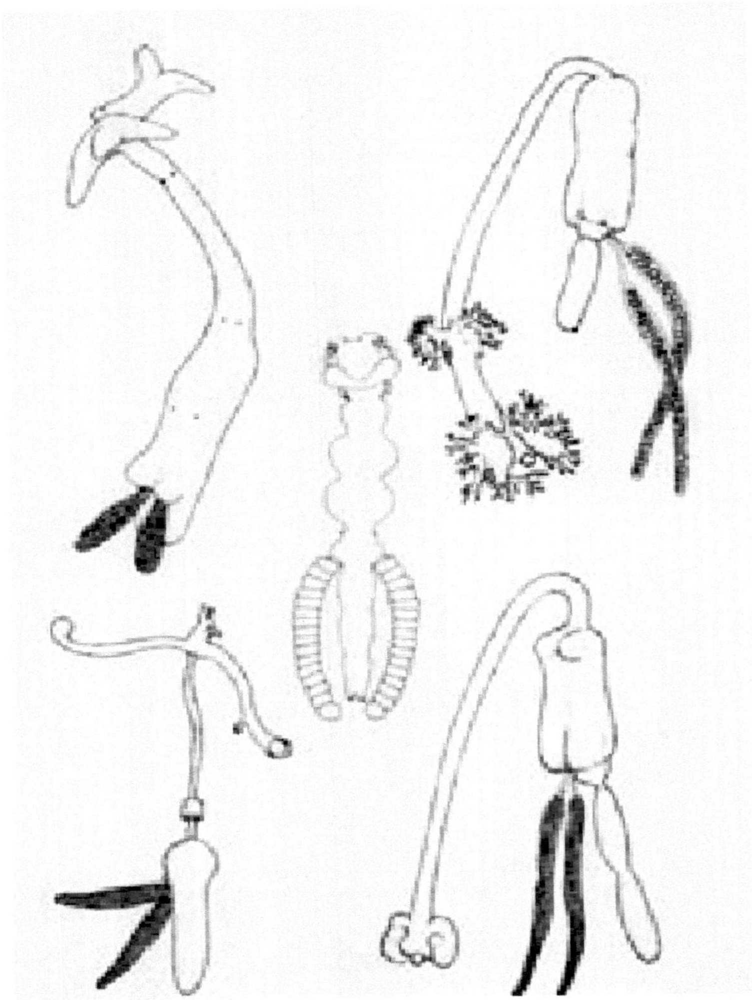

Parasitic Copepods

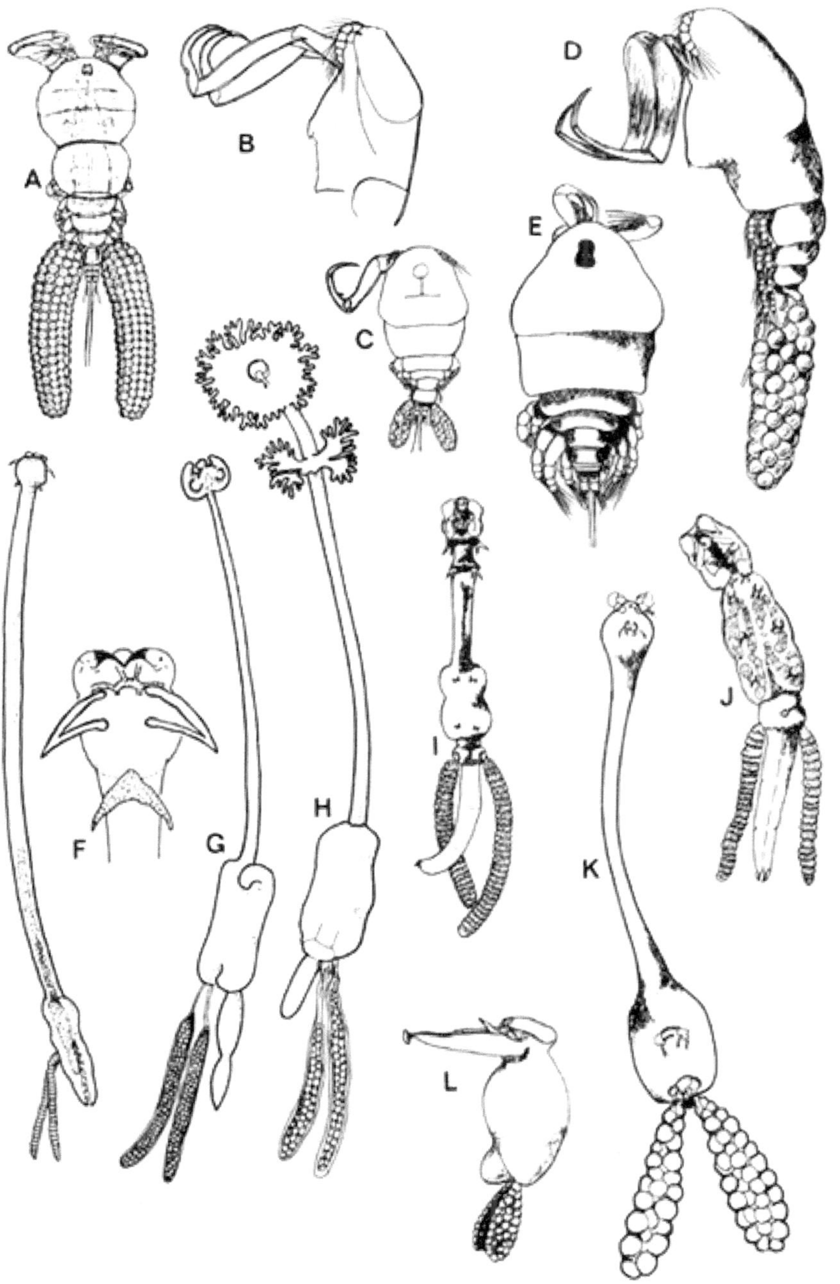

Parasitic Copepoda

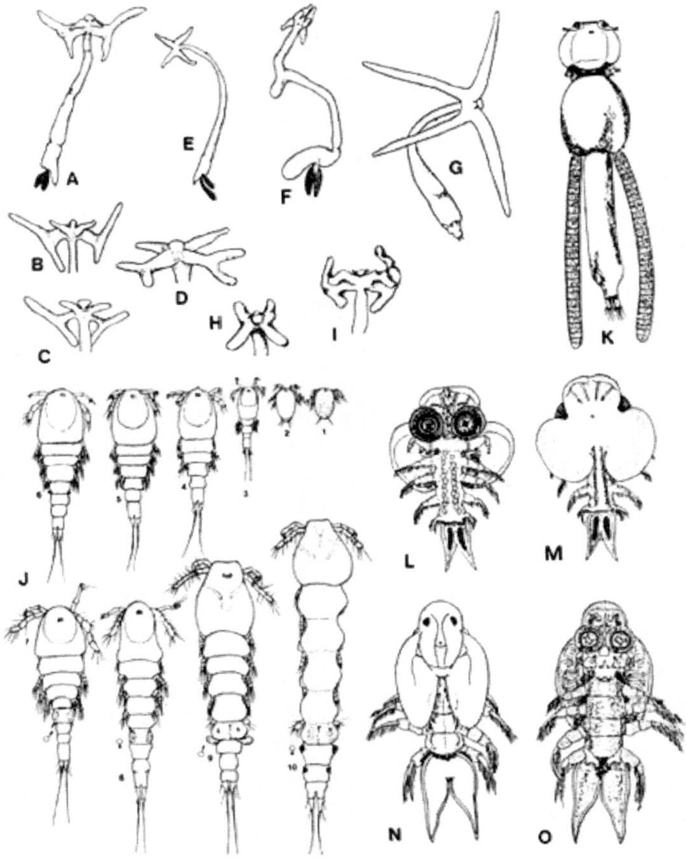

Parasitic Capepoda and Branchiura

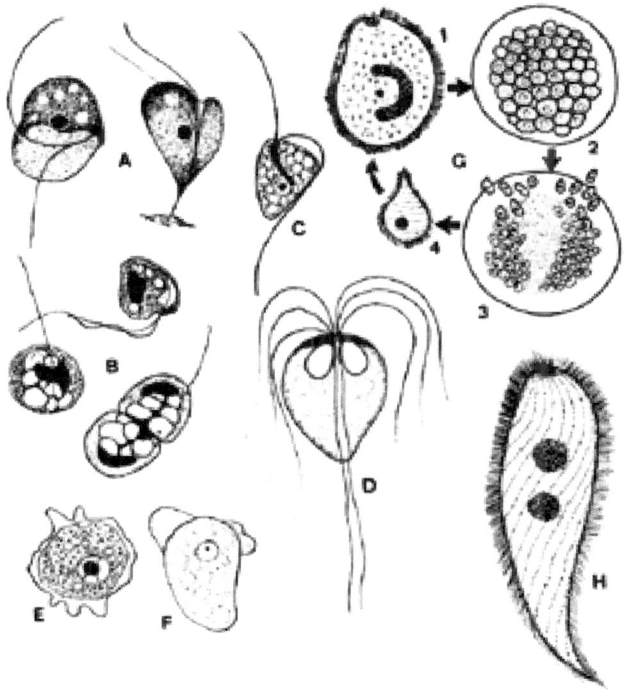

Ectoparasitic and Intestinal Protozoa. A.Ichthyobodo necator free (left) and attached B. *Ichthyobodo* sp. C. *Cruptobia* sp. D. *Hexamita* sp. E. *Thecamoeba* sp. F. *Entamoeba* sp. G. Life cycle of *Ichthyopthirius multiphilis* 1. Trophont 2. Dividing tomont 3. End of division (tomites) theronts escape from the cyst residues 4. Tomites (theoronts) H. *Protoopalina* sp.

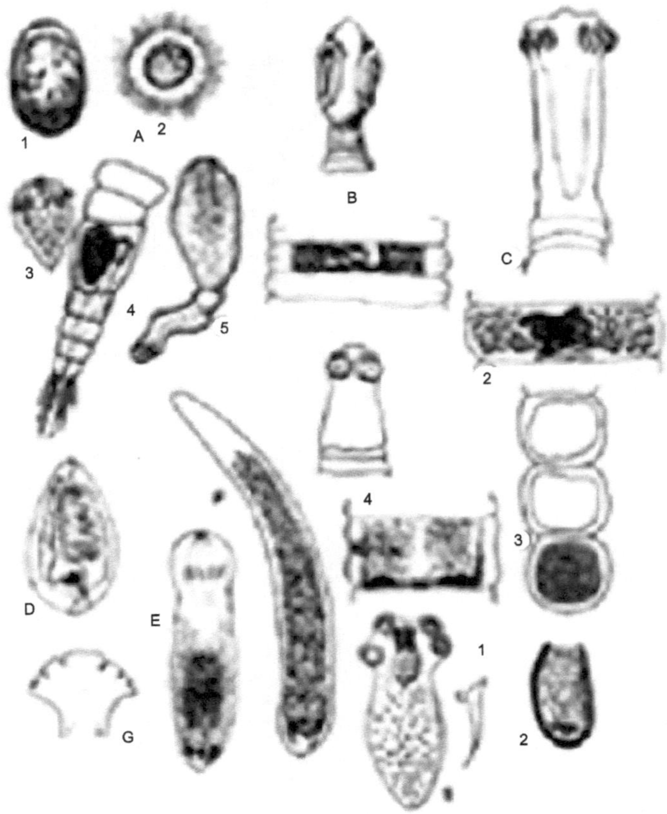

Tape worm Cestoda Bothriocephalus aegypticus A. 1.Coracidia 2. Onchospehere 3. Body cavity of mesocyclops 4. 7 days old procercoid from Mesocyplops, D. scolex C. mature proglotid C. Polyonchobothrim polypteri 1.Scolex 2. Mature proglotid 3.Gravid proglotid D. Amphilina general view E. *Monobothrides woodland* F. *Lytoecystes marcuseni* g. *Proteocephalus largoproglotis* I. *Dilepidid scolices* 1. Free in gut 2. Egg shell in the stomach.

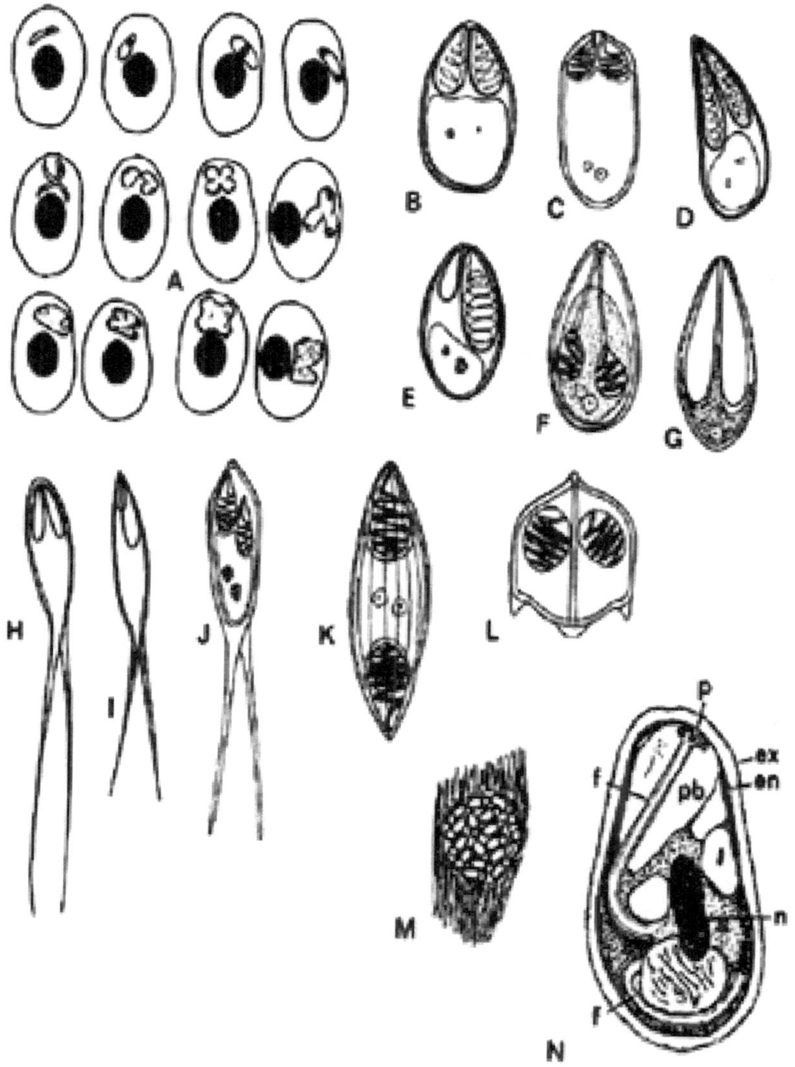

Haemosporida, Microsporia, Microsporia A. Dactyolosoma hannesi B. *Myxobolus* sp. from gill C. *Myxobolus* sp. From viscera D. *Myxobolus sp.* from gill E. *Myxobolus sp.* from gill F. *Myxobolus equatorialis* G. *Myxobolus amieti* H. Henneguya from gills I. Henneguya from gills J. Henneguya from skin K. *Myxidium heterofilamentus* L. *Sphaerospora inequalis* M. *Microspora xenoma* in muscles N. Ultra structure image of a. Microspore spore en: Endospore, Ex: Exospore, f: Filament N: Nucleus P: Polar cap Pb: Polaroblast S: Sporoplasm V: Posterior vacuole.

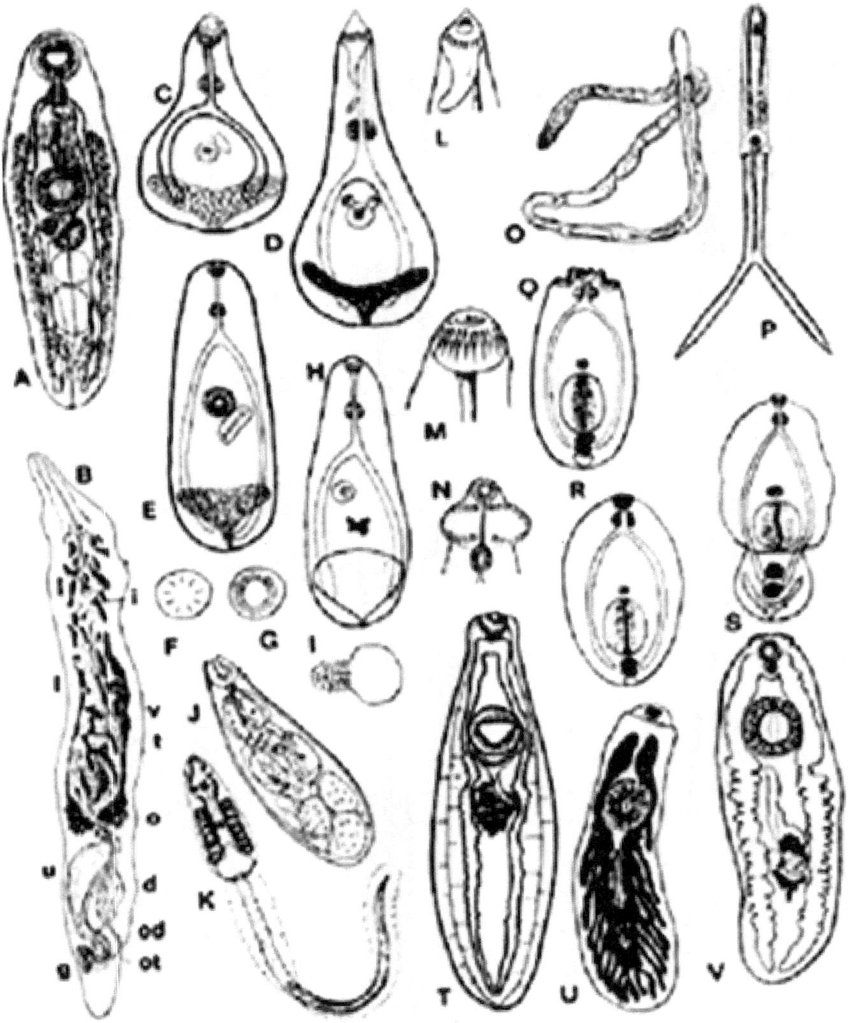

Trematoda A. Allocreadium ghanensis B. Sanguinicola dentata d: Vas deference I: Intestine J. Vagina O: Ovary Od: Oviduct Ot: Ootyp t: Testis U: Uterus V: Vitellaria C. Metacercaria E. Metacercaria of *heterophyes heterophyes* F. Gonotyle of H. aequalis G. Gonotyle of H. hetrophyes H. Met. of stictodiera I.Gonotyle of S. sawakiensis J. Radia of heterophyiid K.Cercaria of heterophyiid. L. Head of Phagicola italica M.Head of Centrocestus sp. N. Head of Echinostomatid metacercaria O. Sporocyst of Diplostomatids P. Furocercaria of Diplostomatids. Q. Diplostomum R & S. Young and developed Neascus (Metacercaria of Diplostumum & *Orinthrodiplostomum* sp.) T. Metacercaria of Clinostomum tilapiae. U. Met. of *Euclinostomum hetrostomum* V. Met, *Clinostomum sp* beneath the scale of Cichlid fish.

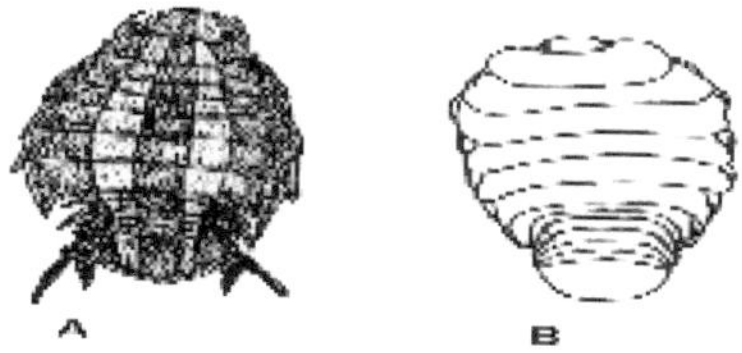

Parasitic Isopoda : A. Nerocila or bignayi, B. *Lironeca tanganyikne*

Ambiphyra sp.

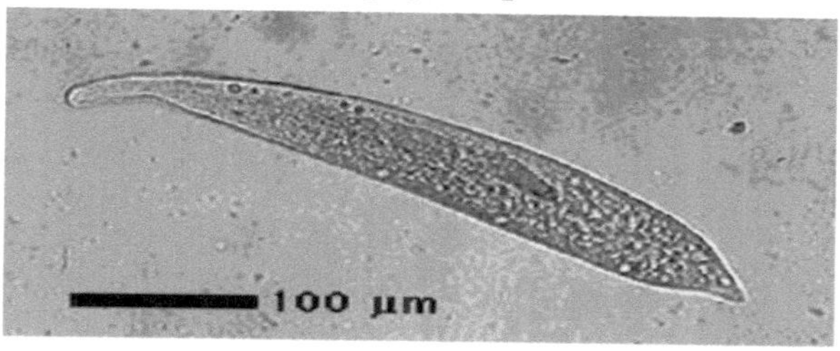

Amphileptus sp.

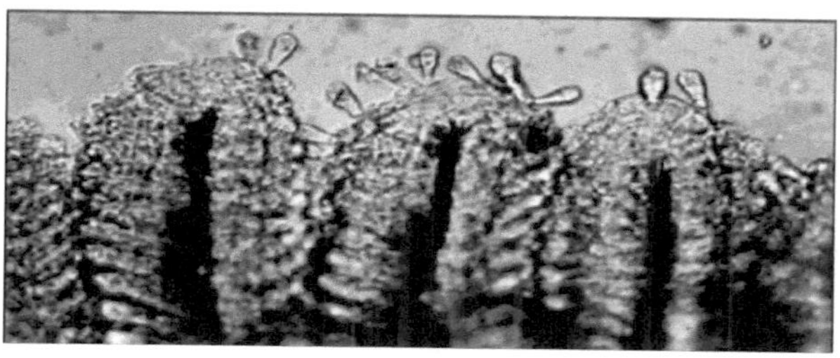

Amphisiella sp.

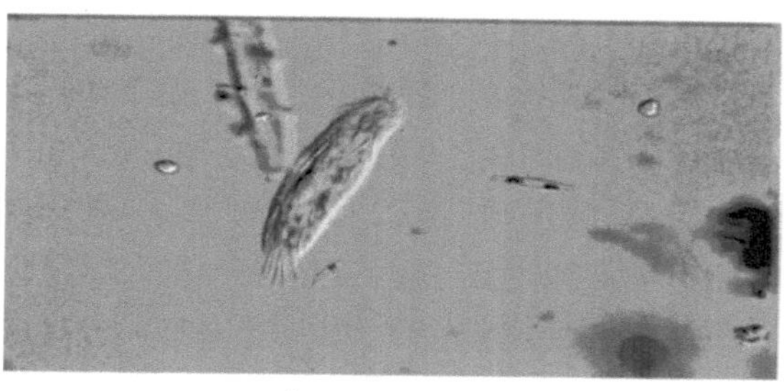

Apiosoma sp.

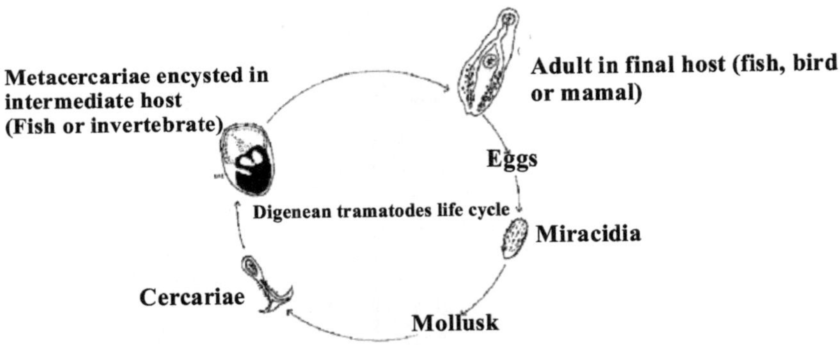

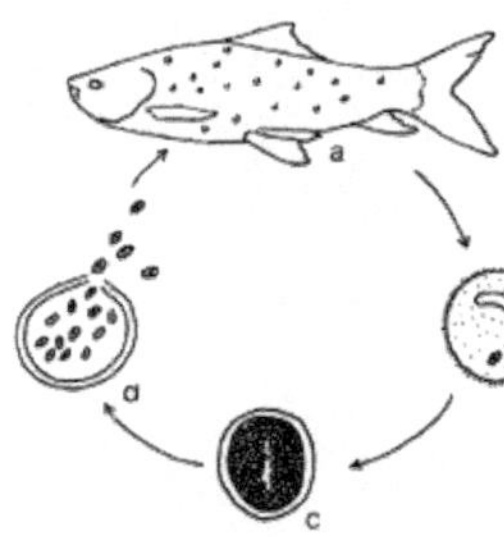

Diagramatic representation of morphology and life cycle of *Ichthypthirius* sp. a. infected fish b. adult parasite C. reproductive cyst D. Infective stage leaving cyst.

Epistylis

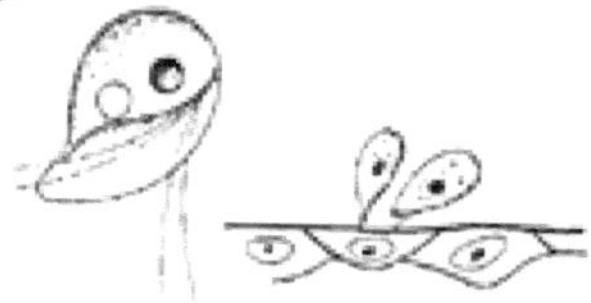

Ichthyoboda

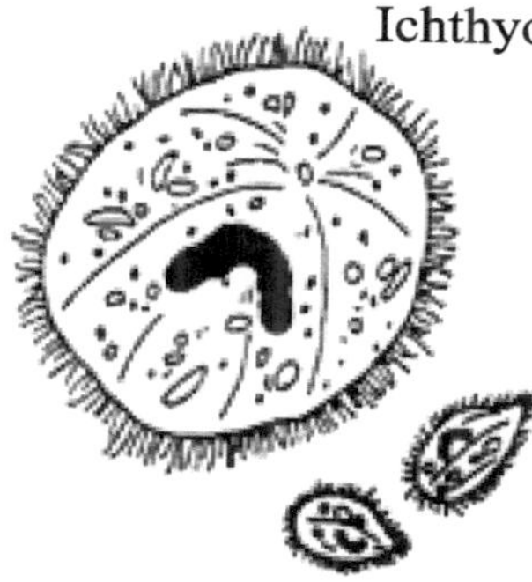

Ich

Henneguya

Hexamita

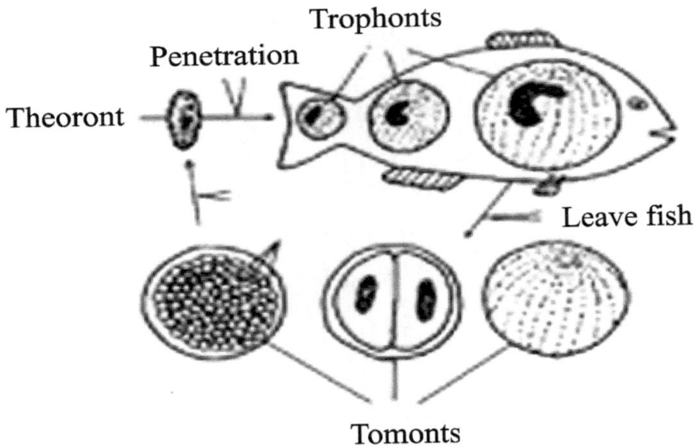

Life cycle of Ich

Probable life cycle of *Ligula sp* causing Ligulosis in Catla catla a- Egg in faeces of bird,b-coracidium larvae c- larval development as procercoid d-Pleurocercoid stage e- final host (bird)

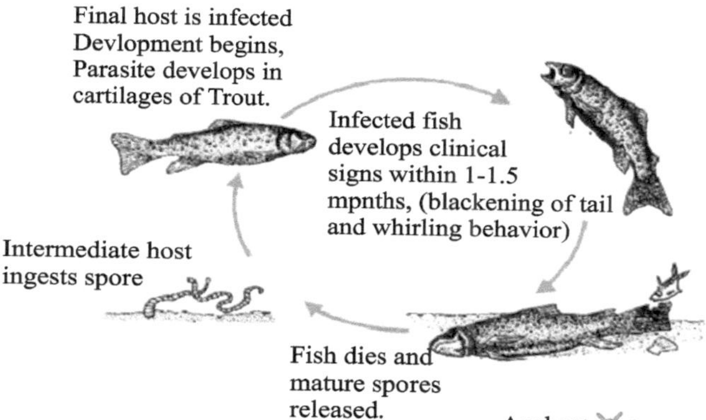

life cycle of Myxobolus

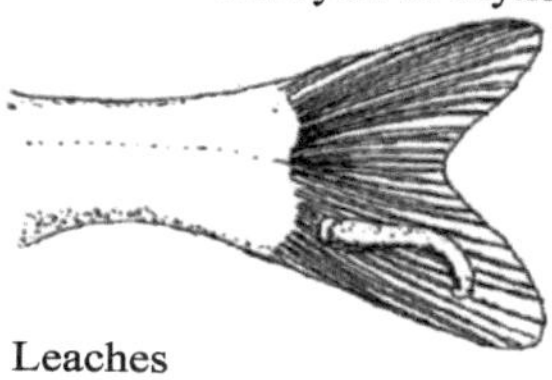

Leaches

Catla catla infected with procercoid larvae of Ligula sp. Pl. Procercoid larvae

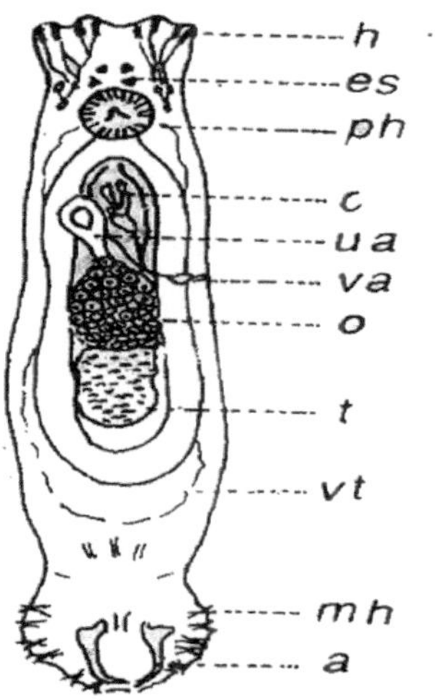

Morphology of an adult *Dactylogyrus sp.* Abreviations: a-anchor c-cirrus es-eye spot h-head organ mh-marginal hooks o-ovary, ph-pharynx, t-tests ua-uterine aperature va- vaginal aperature vt-vitellaria

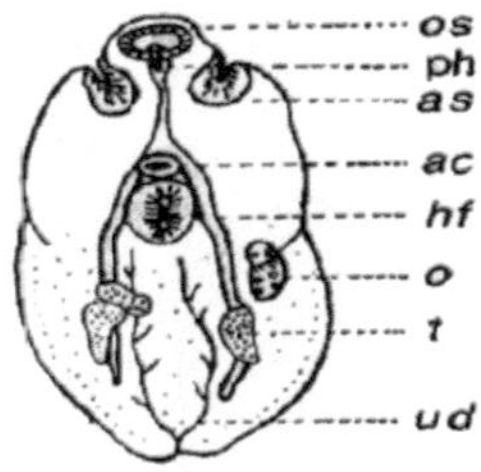

Morphology of a metacercariae of *Diplostomum sp.* ac- acetabulum as-accesory sucker hf- hold fast organ o-Ovary ,os- oral sucker, ph-pharynx t-testes ud-uterine duct.

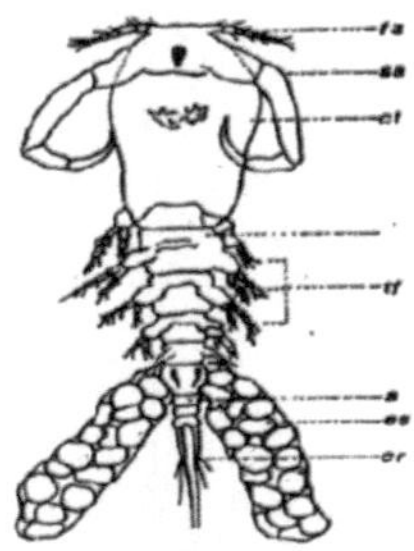

Morphology of adult female *Ergasilus sp.*, Abreviations: a-abdomen ,cr-caudal ramii ,ct-cephalothorax,es-egg saca-first antenna t-thorax , sa-second antenna tf-thoracic feet.

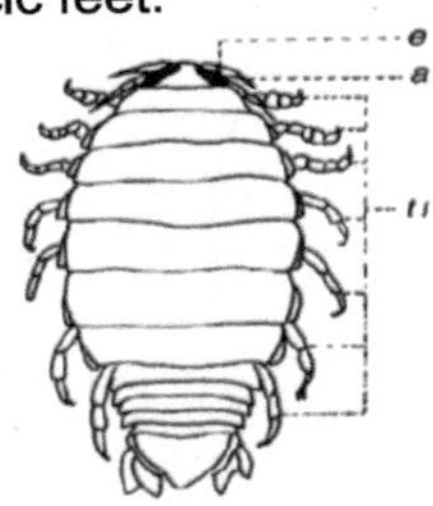

Morphology of adult isopod. A- antenna e- eye tl-thoracic leg

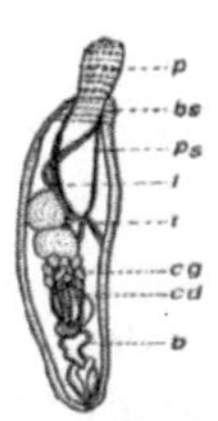

Morphology of an adult male acanthocephalan, b-bursa bs-body spine cd-cement duct , cg-cement gland l-leminisci p-probossis sheath t-testes

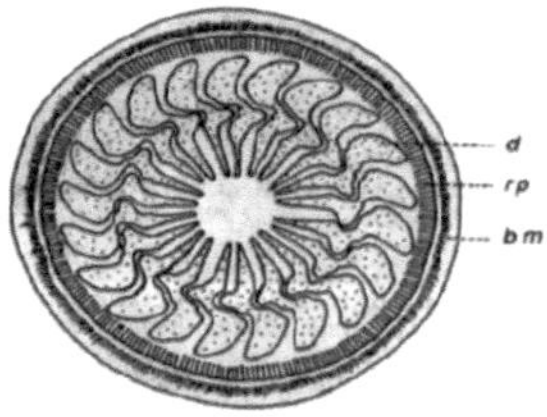

Morphology of *Trichodina sp* bm-border membrane, d-denticle rp-radial pins

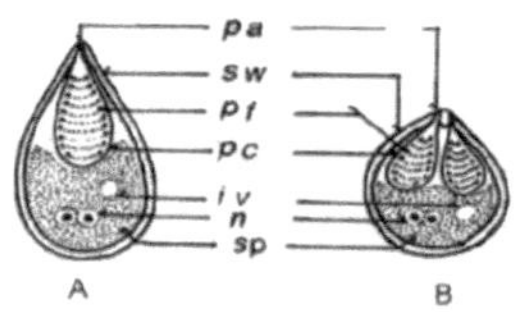

Morphology of Myxozoan spore A- Thelohannellus spore B- Myxobolus spore iv-idiophilus vacuole n-nuclei pa-polar capsule aperature pc-polar capsule pf polar filament sp-sporoplasma sw-spore wall.

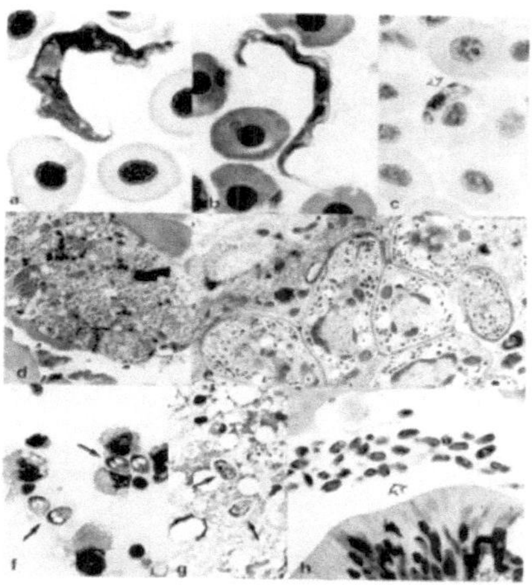

Blood and gut protozoa ,a- trypanosomes in blood.b- a haemogregarine in blood a .c. a light microscopic histology and electronmicroscopic view of a presumed haemogregarine cystozoite stages f. Hexamita sp from tilapia hybrid gut g. *Hexamita sp* from gut of angel fish.h- histlogical section of Hexamita infected Tilapia hybrid.

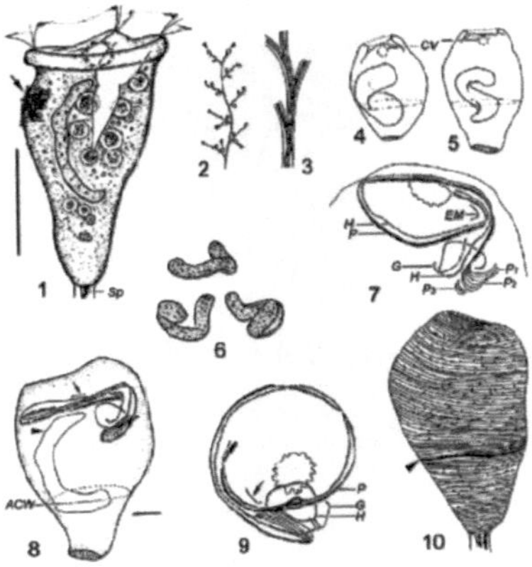

Fig 1-10 Morphology of *Zoothamnium chlamydis sp* n. from life (1-6) and silver impregnation (7-10) 1. Typical Zooid, arrow shows covering of bacteria.2-Colony, arrow heads show the enlarged zooids; 3- talk,note the thick Spasmoneme 4,5- Contracted and enlarged zooids.6- Twisted Macronucleus in enlarged zooids.7,9- Oral apparatus Arrow in 9 indicates epistomal membrane 8. General infraciliture arrow marks epistomial membrane.arrowhead shows macronucleus, aboral ciliarry wreth ;10- silverline system, arrow head shows aboral ciliary wreth ACW- aboral ciliary wreth CV; Contractile vacuole EM; Epistomal membrane G; Germinal kinety H; Haplokinety; P- Polykinety p-1-3 Peniculus 1-3; S- Spasmoneme

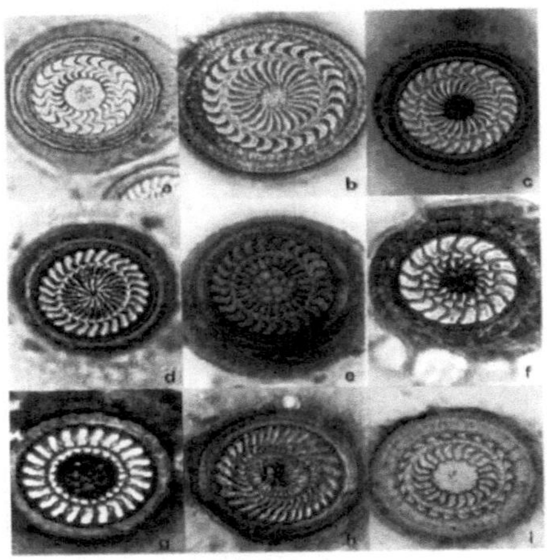

Ectoparasitic Protozoa, Tricodinids, (Silver impregnated) a. Tricodina compacta (O. mosambicus) b. T. mrigala (O. mosambicus), c. T. mutabilis Barbus sp. d. T. centrostriata. (O. mosambicus) e. T. reticulate, Farmed gold Fish. f. T. minuta (O. mosambicus) g. Trichodiniella epizootica (Carp) h. Tripartilla cichilidarum (O. mosambicus) i. Hemitrichodina robusta (O. Mosambicus)

Nematode

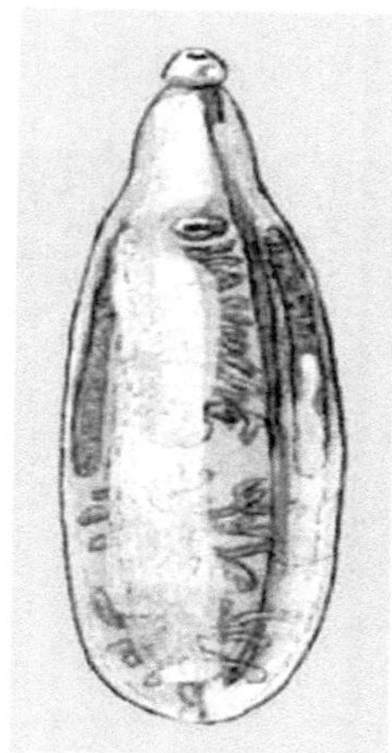

Chlonorchnis simensis

Juvenile Camallanus nematode in Cichilid

Blepharisma

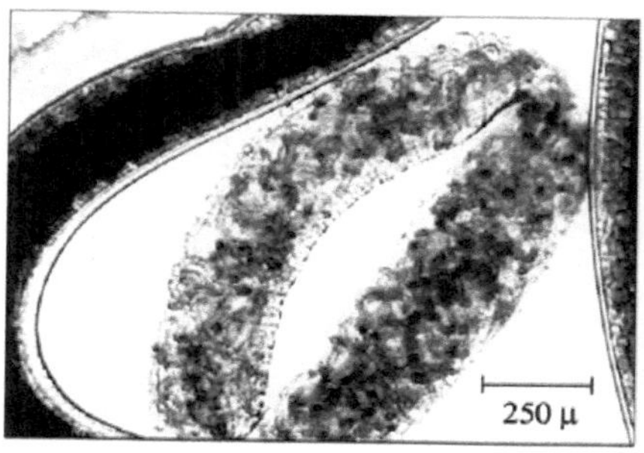

Close of *Camallanus sp* female showing larvae inside

Ambhibia

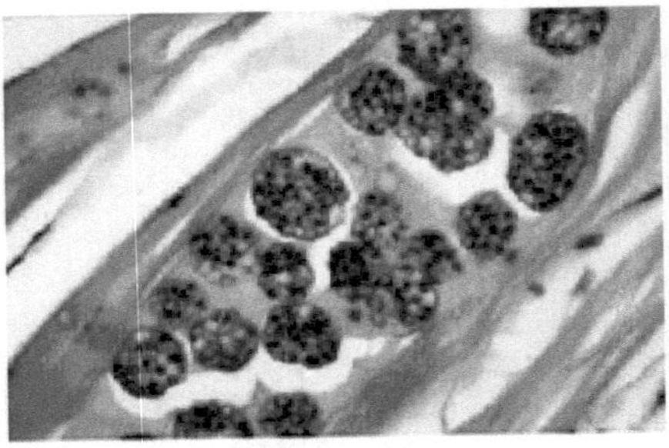

Pleistospora sp

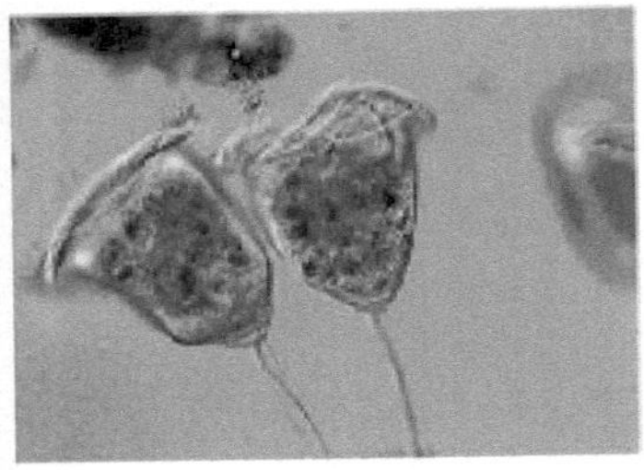

Vorticella sp.

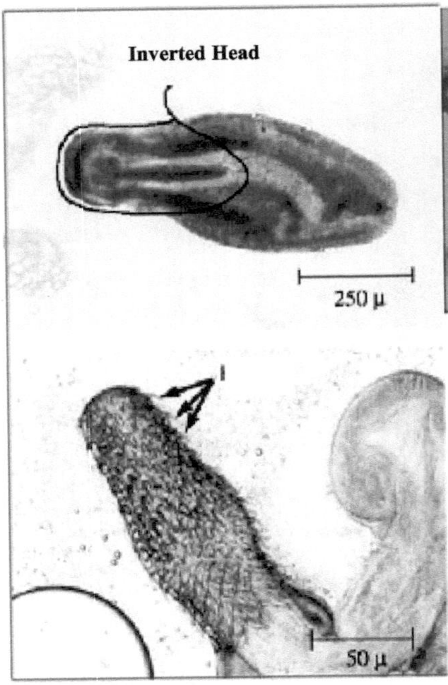

Acanthocephalan with head inverted (top) and head averted showing typical thorny head hooks (bottom)

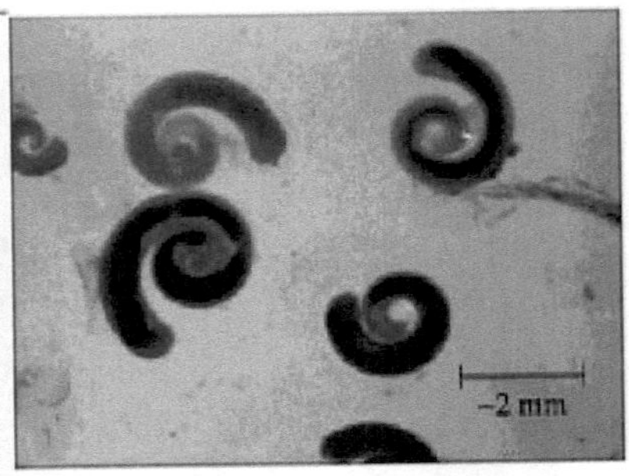

Pentastomes showing typical shape

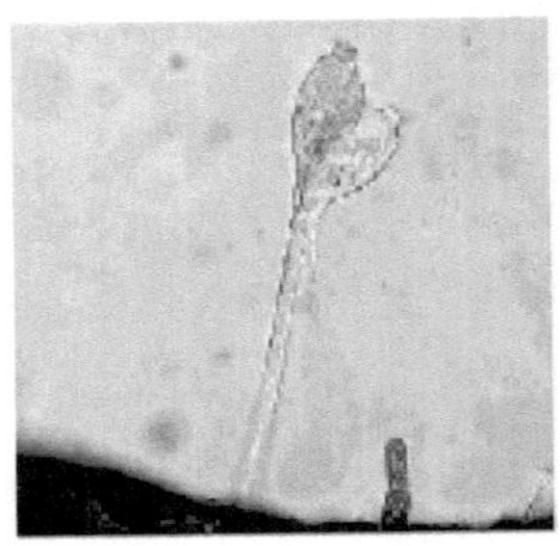

Epistylis sp.

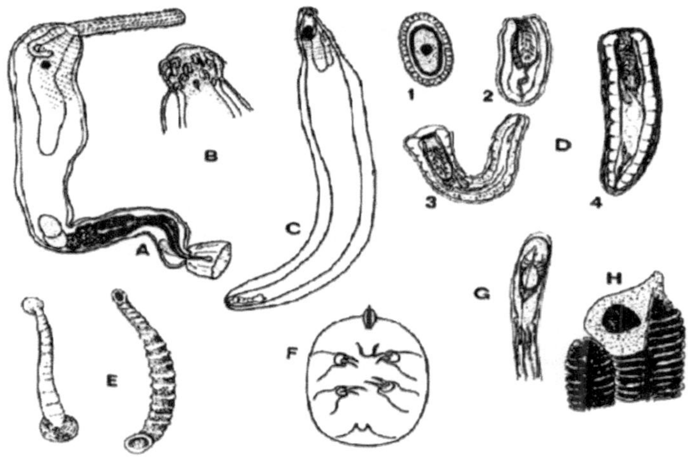

Acanthocephalan and various other parasitesof fish. A. Paragorgorhynchusschariensis, male B.C. *Palicentis Tetradontis*, female D. Larval stages of Acanthocephalans 1. Egg of Neoechinorrhynchus 2. Acanthela from Gamarus amphipod . 3. Acathella from Ostracods. 4. Cystcanthus(Acanthor) from fish E. Piscicolid leeches (Hiridinea) Pentastomid larva G. Parasitic larva of mutilid bivalve H. Unionid glochidium embded in the gill tissue.

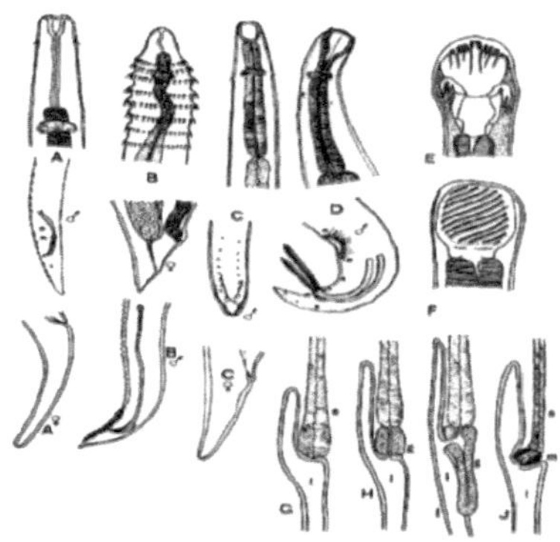

Nematoda , A. Rhabdocona congolensis , Posterior and Anterior ends .B. Spinitectus allaeri anterior and.posterior ends, C. Procamalanus laevionchus anterior and posterior ends.D. *Cucullanus barbi*, male, anterior and posterior ends E. *Paracamallanus cyathopharynx* mouth capsule F. *Spirocamallanus spiralis* mouth capsule. G. Position of Ventricular (g) and intrstinal (I) coecca in larval Heterocheilidae G. Amplicaecum , H. Porrocaecum I.Contrcaecum J. Dujardinascaris with muscular ventriculus (m), O. Oesophagus.

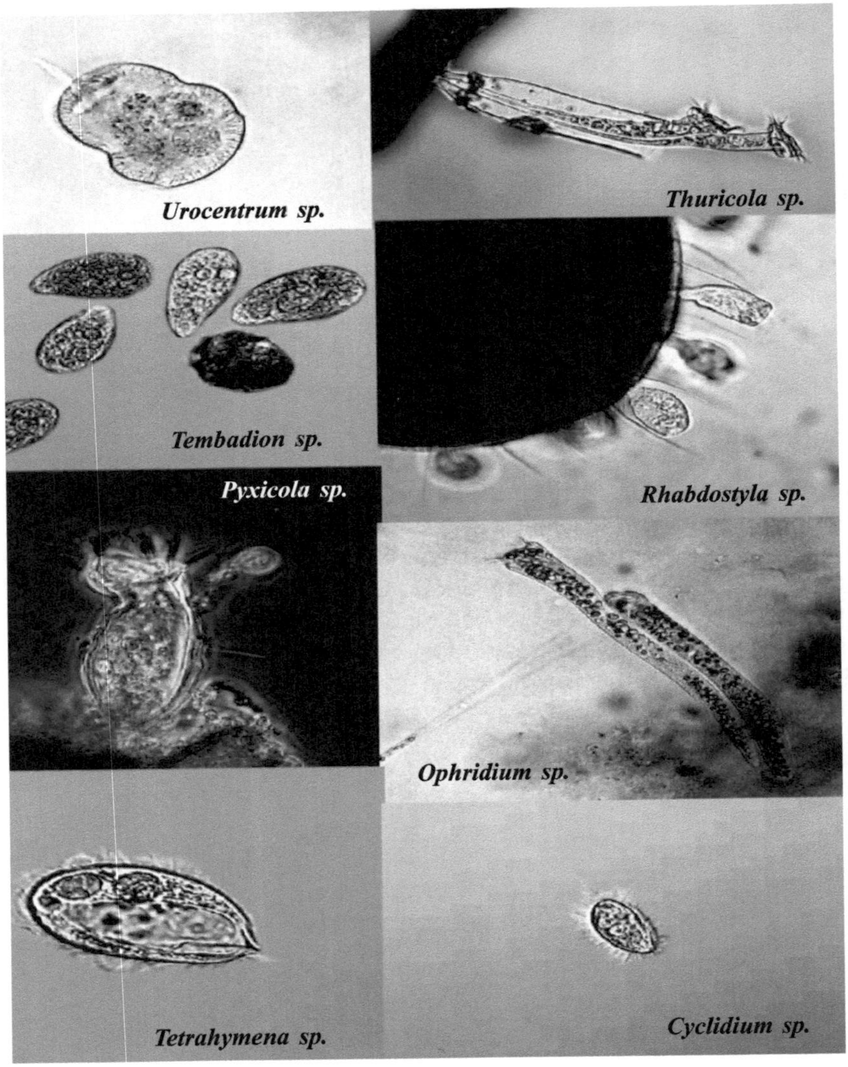

Urocentrum sp. *Thuricola sp.*

Tembadion sp. *Rhabdostyla sp.*

Pyxicola sp. *Ophridium sp.*

Tetrahymena sp. *Cyclidium sp.*

❑❑❑

Suggested Literatures

Acha P.N. & Szyfres B. 2003. Zoonoses and communicable diseases common to man and animals, 3rd Ed., Vol. 3, *Parasitoses*. Pan American Health Organization, Washinton.

Adamczyk, L. H. 1981. Parasite fauna of carp fry (*Cyprinus carpio*) bred in ponds fed by rainwater. *Wiadomosci-Parazytologiczne,* **27**(6) : 763-722.

Ahmed, A. ALI, S.M.K Samad, A. (1991) Probable cause of fish culture in Bangladesh. Nutrition news. **14**(1): 3. (p 6)

Ahmed, A. T. A. and Begum, R. 1978. Study on the distribution of some endoparasitic helminths in six fresh water fishes of Dhacca and Barisal. *Bangladesh J. Aquac.*, **1(1)** : 52 – 60.

Ahmed, A. T. A. and Ezaz, M. T. 1997. Diversity of helminth parasites in the freshwater catfish of Bangladesh. *Dis. Asian Aquacult.* III. pp. 155- 160

Ahmed, A. T. A. and Ezaz, M. T. 1997. Diversity of helminth parasites in the fresh water catfish of Bangladesh. *Dis. Asian Aquacult.*, **111. p.155 – 160.**

Ali, N. M.; Abul-Eis and Abdul-Ameer, K. N. 1988. Study on parasites of common carp *Cyprinus carpio* and other freshwater fishes in Habbaniyah Lake, Iraq. *J. Biol. Sci. Res.,* **19**(2) : 395-408.

Alikunhi, K. H. 1957. Fish culture in India. Farm Bulletin of Indian Council of Agricultural Research, No. **20**, pp-1 - 150.

Al-Marzouq, A. and Al-Rifae, K. 1994. *Benedenia sp.*, a monogenetic parasite of cultured brown spotted grouper, *Epinephelus tauvina*, in Kuwait. *J. Aqua. Trop.* **9**: 255-258.

Andersen, R. M. 1974. Population dynamics of the cestoda *Caryophyllaeus laticeps* (Pallas, 1781) in the bream (*Abramis brama* L.). *J. Animal Ecol.*, **43**: 305- 321.

Andersen, R. M. 1976. Seasonal variations the population dynamics of *Caryophyllaeus laticeps*. *Parasitology,* **72**: 281- 305.

Anderson, D. P., 1974. Fish immunology. In: Diseases of fishes Stanisias. F. Sneiszko. F. & Axelrod (eds.) H. R. F. H. publication Inc. New Jersey 07753 USA

Anderson, J. I. W. and Conroy, D. A. 1992. The pathogenic myxobacteria with special reference to fish disease. *J. Appl. Bacteriol.,* **32** (30): 9.

Anderson, R. C. 1984. The origin of zooparasitic nematodes. *Can. J. Zool.* **62**: 317- 328.

Anderson, R.C.1992. Nematode Parasite of Vertebrates : Their development and transmission. CAB International, 578 pp.

Andrade-salas, O., Pineda-Lopez, R. E. and Garcia-Magana. 1994. Spirocamallanus rebrcae sp. n. (Nematoda: Camallanidae) from freshwater fishes in South-eastern Mexico. *Folia Parasitilogica.*, 41: 259- 270.

Andrews, C., Chubb, J. C., Coles, T. and Dearsley, A. 1981. The occurrence of *Bothriocephalus acheilognathi Yamaguti*, 1934 (B. gowkongensis) (Cestoda: Pseudophyllidea) in the British Isles. *J. Fish Dis.* **4**: 8993.

Aohagi, Y., Shibahara, T., Machida, N., Yamaga, Y. and Kagota, K. 1992. *Clinostomum complantum* (Trematoma : Clinostomatidae) in five new fish hosts in Japan. *J. Wildl. Dis.*, **28** (3): 467- 469.

APHA (American Public Health Association) American Water Works Association and Water pollution Control Federation, 1998. Standard methods for the examination of water and wastewater. 20th edition, Am. Publ. Health. Assoc., Washington, USA, pp: 10-1-60-61.

Arnott, S. A., Barber, I. and Huntingford, F. A. 2000. Parasite – associated growth enhancement in a fish – cestode system. *Royal Soc. London. Proc. Biol. Sc.* **267**(1444):657- 663.

Arnott, S.A., Barber, I., and Huntingford, F.A., 2000, Parasite-associated growth enhancement in a fish-cestode system, Proceedings of the Royal Society of London B **267:** 657-663.

Atamanalp, M., Kocaman, E. M. and Canyurt, M. A. 2002. The effects of sewage on haematocrit and sediment of *C. capoeta. Su urunleri dergisi. J. Fish. Aquat. Sci.*, **19** (3-4) : 439-445.

Avault, J. W. 2000. Research report – A new parasite for channel catfish. *Aquacult. Mag.* **26**(5) : 48- 49.

Avenant, A. and Van As, J. G 1986. Observations on the seasonal occurrence of the fish ectoparasite *Dolops ranarum* (stuhlmann, 1891) Crustacea: Branchiura) in the Transvall. *S. Afr. J. Wildl. Res.*, **16** : 62-64.

Awa, J., Anyanwu, P. and Ezenwa, B. 1988. Incidence of parasitic infection of pond raised Tilapia spp. And some cultivable fish species from three ecological areas of Lagos State. *Tech. Pap. Niger. Inst. Oceanogr. Mar. Res.*, **32**: 24.

Awachie, J. B. E. 1966. The development and life history of *Echinorhynchus truttae* Schrank, 1788 (Acanthocephala). *J. Helminthol.* **40**: 11- 32.

Aydogdu, A. and Altunel, F. N. 2002. Helminth parasites (Platyhelminthes) of common carp (*Cyprinus carpio*) in Iznik Lake. *Bull. Eur. Assoc. Fish Pathol.*, **22 (5)** : 343-348.

Ayyappan, S. and Diwan, A. D. 2006. Fisheries research and development in India. *Fishing chimes*, **26** (1): 19- 23.

Ayyappan, S., 1987. Investigations on the limnology and microbial ecology of a benthic habitat. Ph.D. Thesis, Bangalore University, Bangalore, 330pp.

Balakhnin, I. A.; Guri-Kovskii, S. A. and Kosolap, M. M. 1993. Population numbers and distribution of ectoparasites in carp. *Parasitologiya,* **27**(2):140-147.

Balakhnin, I. A. and Kurovskaya, L. Ya. 1996. Results of morphophysiological analysis of one year carps of different sizes invaded by Bothriocephalus. *Dopov. Akad. Nauk Ukr./Dokl. Akad. Nauk Ukr.,* **2** : 146-148.

Balm, P.H.M.,Carrick T.R., Mcoenen A.J., Pottinger T.G,1996.Trychophyra intermedia on the gills of rain bow trout acclimating to low ambient pH,*Journal of fish biology*. **Vol 48,** 147-150.

Banerjee, S. M. 1967. Water quality and soil condition of fishponds in some states of India in relation to fish production. *Indian J. Fish.,* **14** (1 & 2): 115- 144.

Barber, I., and Huntingford, F.A., 1995, The effect of *Schistocephalus solidus* (Cestoda: Pseudophyllidae) on the foraging and shoaling behaviour of three-spined sticklebacks, *Gasterosteus aculeatus*, **Behaviour 132,** 1223-1240.

Barber, I., and Huntingford, F.A., 1996, Parasite infection alters schooling behaviour: deviant positioning of helminth-infected minnows in conspecific groups, Proceedings of the Royal Society of London **B 263:** 1095-1102.

Barber, I., Downey, L.C., and Braithwaite, V.A., 1998, Parasitism, oddity, and the mechanism of shoal choice, Journal of Fish Biology 53: 1365-1368.

Barber, I., Hoare, D., and Krause, J., 2000, Effects of parasites on fish behaviour: a review and evolutionary perspective, Reviews in Fish Biology and Fisheries **10:** 131-165.

Barber, I., Huntingford, F.A., and Crompton, D.W.T., 1995, The effect of hunger and cestode parasitism on the shoaling decisions of small freshwater fish, Journal of Fish Biology **47:** 524-536.

Bardach, J. E., Ryther, J. H. and McLearney, W. O. 1971. Aquaculture. Wiley- Interscience, New York, 868 pp.

Bartholomew, J.L., Rohovec, J. S.and Fryer, J. L.1989. Ceratomyxa shasta, a Myxosporean Parasite of Salmonids (LSC Fish Disease Leaflet 80)

Basu, S. P. 1950. Myxosporidian parasites of Indian major carps, *Catla catla* (Ham.), *Labeo rohita* (Ham.), *Cirrihinus mrigala* (Ham), *Labeo bata* (Ham.) and their seasonal variations. *Proc. Indian. Sci. Congr.* , **37** : 239.

Bauer, O. N. and Hoffman, G. L. 1976. Helminth range extension by translocation of fish. In: Page, L. A. (ed.), Wildlife diseases, Plenum Publishing Corp, N.Y. pp. 163- 172.

Bauer, O. N., Musselius, V. A., Strelkov, J. A. (1981). Diseases of pond fishes. Publishing House Legkhaya i Pishtsevaya Promishlennost, Moscow (in Russian)

Baylis, H. A. 1923. Notes on *Procamallanus spiralis* Baylis, 1923 (Nematoda). *Parasitologiya.* **15**(2): 137- 138.

Bazal, K., Lucky, Z. and Dyk, V. 1969. Localization of fish-lice and leeches on carps during the autumn fishing. *Acta Veterinaria (Brno).*, **38** : 533-544.

Bhowmik, M. L., Pandey, B. K. and Sarkar, U. K. 1994. Microbial and chemical changes in water and sediment of an experimental wastewater fed fishpond. Paul Raj, S. (Ed). *Proc. Nat. Symp. Aquacult. for 2000 AD.* , pp : 323-328.

Biswas, A. 2002. Spontaneous EUS of freshwater fishes: Prevalence, isolation, clinical signs and pathomorphological studies. *M.V.Sc. thesis*, submitted to W.B.U.A.F.S., Kolkata, India.

Blaxhall, P. C. and Daisley, K. W. 1973. Routine haematological methods for use fish with blood. *J. Fish Biol.* , **5** : 771-781.

Bose, K. and Sinha, A. K. 1983. Histopathology of Clarias batrachus (Linn.) attributable to the intestinal cestode, *Lytocestus indicus* Moghe, 1925. *Proc. Nat. Acad. Sci. India,* **53**: 226-230

Bose, P., Noor Islam S.K.and Dash ,G. (2009). Prevalence of ornamental fish diseases in relation to water quality parameters of Howrah district, West Bengal J.Adv. Zool. **30(1):** 34-43

Bower-shore, C. 1940. An investigation of the common fish louse, *Argulus foliaceus* (Linn.). *Parasitology.* , **32** : 361-371.

Boyd, C. E. 1979a. Water quality in warm water fish ponds. Auburn University Agricultural Experiment Station, Auburn, Alabama, 359 pp.

Boyd, C. E. 1982. Water quality management for pond fish culture. Elsevier Scientific Publishing Company, Oxford, New York. pp. 1- 113, 318.

Boyd, C. E. 1990. Water quality in ponds for Aquaculture, Birmingham Publishing Co., Alabama 482 pp.

Brown, D. and Brooks, A. 2002. A Survey of Disease Impact and Awareness in Pond Aquaculture in Bangladesh, the Fisheries Training and Extension Project – Phase II. pp. 85- 93.

Brown, E. E. and Gratzek, J. B. 1983. Fish farming handbook: food, bait, tropicals and goldfish. AVI Publishing Company, Inc. Westport, Connecticut. 391 pp.

Buchmann, K. and Bresciani, J. 1997. Parasitic infections in pond-reared rainbow trout *Oncorhychus mykiss* in Denmark. *Dis. Aqua. Orga.*, **28** (2) : 125-138.

Buchmann, K., Koie, M and Prento, P. 1987a. The nutrition of gill parasitic monogenean Pseudodactylogyrus anguillae. *Parasitol. Res.*, **73:** 532-537.

Buchmann, K., Mellergaard, S. and Koie, M. 1987b. *Pseudodactylogyrus* infections in eel; a review. *Dis. Aquat. Organisms.* **3**: 532- 537.

Cable, J., Scott, F. C. G., Tinsley, R. C. and Harris, P. D. 2002. Behaviour favouring transmission in the viviparous monogenean *Gyrodactylus turnbulli. J. Parasitol.* **88**(1): 183- 184.

Cali, A. 1991. General microsporidian features and recent findings on AIDS isolates. *J. Protozool.* **38:** 625-630.

Carvajal, J., Uribe, J. C. and Vial, M. V. 1991. Parasitos estresantes amenazan salmons en cultivo. *Creces.*, **6-7:** 20- 22.

Carvalho, L.N; Del-Claro, K and Takemoto, R. M. 2003. Host-Parasite interaction between brachiurans (Crustacea: Argulidae) and piranhas (Osteichthyes: Serrasalminae) in the Pantanal wetland of Brazil. *Environ. Biol. Fish.*, **67** (3) : 289-296.

Casack, R. 1986. Development of infections of *Gyrodactylus colemensis* Miizelle and Kritsky, 1967 (Monogenea) and their effect on fry of salmo gairdneri Richardson. *J. parasitol.* **72** : 663- 668.

Chakrabarty, A. N. and Dastidar, S. G. 1991. Repeated isolation of chemoautotrophic nocardioform bacteria from fish epizootic ulcerative syndrome. *Indian J. Expt. Biol.,* **29**: L623- 627.

Chanratchakool, P., Turnbull, J. F. and Limsuwan, C. 1994. Health management in fish ponds. *Aquatic Animal Health Research Institute*, Thailand, pp. 91.

Chappell, L. H. 1969. The parasites of the three spined stickleback *Gasterosteus aculeatus* L. from a Yorkshire pond. II. Variation of the parasite fauna with sex and size of fish. *J. Fish. Biol.* **1**: 339- 347.

Chattopadhyay, G. N., Saha, P. K and Ghosh, A. 1989. *Proc.Nat. Sem. Freshwater Aqua.*, pp. 45-46.

Cheruvat, D., Bharathan, C. K., Namboodiri, N. N. and Sidharthan, T. 2002. Argulosis infection in carps and its control strategies: Preliminary observations. *Fish. Chimes.* , **21** (10-11) : 34-100.

Ching, H. L., and D. R. Munday. 1984a. Geographic and seasonal distribution of the infectious stage of Ceratomyxa shasta Noble, 1950, a myxozoan salmonid pathogen in the Fraser River system. *Can. J. Zool.* **62:**10751080

Choon Shil Lee, (2009). *Korean J. Parasitol*. Vol. 47, Supplement: S103-S113, October.

Chowdhury, M. M. 2002. Argulosis- a Spreading Disease in Bangladesh. Department of Fisheries, University of Dhaka. In *www.google.co.in.*

Chubb, J. C. 1979. Seasonal occurrence of helminths in freshwater fishes. Part II Trematoda. *Adv. Parasitol.,* **17** : 141-313.

Cone, D. K. 1995. Monogenea (Phylum Platyhelminthes). Woo. P.T.K. (ed.), Fish diseases and disorders, Vol. I. Protozoan and Metazoan Infections.

Conroy, C. V. and Herman, R. L. 1997. Crustacean Parasites of Fish, *A Text Book of Fish Diseases*, Narendra Publishing House, Delhi. , PP: 233 - 241.

Cross, J. H. 1990. Intestinal capillariasis. *Parasitol Today.*, **6 :** 26-28.

Cross, J. H. and Basaca-Sevilla, V. 1991. *Capillaria philippinensis* : a fish-born parasitic zoonosis. *Southeast Asian J. of Trop. Med. and Pub. Health.*, **22** (suppl) : 153-157.

Crowden, A.E., and Broom, D.M., 1980, Effects of the eyefluke, *Diplostomum spathaceum*, on the behaviour of dace (*Leuciscus leuciscus*), Animal Behaviour **28:** 287-294.

Cunningham, E.J., Tierney, J.F., and Huntingford, F.A., 1994, Effects of the cestode *Schistocephalus solidus* on food intake and foraging decisions in the three-spined sticklebacks *Gasterosteus aculeatus*, Ethology **79:** 65-75.

Current, W.L. & Janovy, J. Jr., 1976. Ultrastructure of interlamellar *Henneguya exilis* in the channel catfish. *J. Parasitol.,* **62:** 975–981.

Das, B. K.1998. Studies on the effect of some pesticides and commonly used chemicals on Indian Major Carps and their ecosystem. *Ph.D. Thesis.* Submitted to O.U A.T. Bhubaneswar. Orissa. pp: 1- 142.

Das, M. K. 1994. Outbreak of the fish disease, epizootic ulcerative syndrome in India- an overview. **In** R.J. Roberts, B. Campbell and I.H. MacRae, (Eds) ODA *Regional Seminar on Epizootic Ulcerative Syndrome, at the Aquatic Animal Health Research Institute,* Bangkok, Thailand, 25-27 January 1994. , pp : 21-35.

Das, M. K. and Das, R. K. 1995a. Fish Diseases in India- A Review. *Environ. & Ecol.* , **13 (3) :** 533-541.

Das, M. K. and Das, R. K. 1997. Fish and prawn disease in India - Diagnosis and control. *Inland Fisheries Society of India, CIFRI. Barrackpore,* 139 pp.

Das, M. K. and Das, R. K. 1997. Fish and Prawn diseases in India - Diagnosis and control. J. Inland Fish. Soc. India. , **20: 133**

Das, M. K. and Haldar, D. P. 1986. Studies on protozoan parasites from fishes 2 new species of the genus *Neothelohanellus* n. Gen. *Arch. Protistenk.,* **132**(1): 125- 130.

Das, M. K. and Pal, R. N. 1987. Histopathology of gill infestation by monogenea and urceolariid ciliates in carps cultured in India. *Ind. J. Parasitol.,* **11**(2): 127- 130.

Das, M. K., Das, R. K. and Mondal, S. K. 2002. Some stress sensitive parameters of young major carp, *Labeo rohita* (Hamilton-Buchanan). , *Indian J. Fish.* , **49** (1) : 73-78.

Das, M. K., Halder, D. P. 1987. Ulceolariid ciliates of the genus Tripartiella invading gills of freshwater cultured carps in India. *Arch. Protistenk.,* **134**: 169- 178.

Das, M. K., Pal, R. N. 1987. Histopathology of gill infestation by monogenea and urceolariid ciliates in carps cultured in India. *Ind. J. Parasitol.,* **11**(2) : 127- 130.

Das, P. C., Ayyappan, S. and Jena, J. K. 2006. Haematological changes in the three Indian major carps, *Catla catla* (Hamilton), *Labeo rohita* (Hamilton) and *Cirrhinus mrigala* (Hamilton) exposed to acidic and alkaline water pH. *Aquaculture.* , **256** (1-4) : 80-87.

Das, R. 2003. Study of helminth parasites in the catfishes of commercial importance from Nadia district of West Bengal. *M.F.Sc. Thesis* submitted to West Bengal University of Animal and Fishery Sciences. pp. 121.

Das, R. and Cone, D. K. 1986. A review of parasites as vectors of viral and bacterial diseases of fish. *J. Fish. Dis.,* **9** : 169-171.

Das, R., Dash, G. and Mondal, S. (2006). Prevalence of Nematodes in cat fishes of commercial importance from Nadia District, West Bengal. *Aquacult.* **7(2)** : 327-330.

Dash G. and Chatterjee,N.R.(2000).Strategy for Environmental Management in Aquaculture. *Indian J.Environ.& Ecoplan.,* **3(3):** 749-752.

Dash, G. and Bandopadhya., P.K.(2000). The Causative Factors of EUS in India, *Geobios,* **27 (2-3)** : 95-97

Dash, G. and Behera, S. (1999). Ozone based treatment of water and waste water, *Green technology.* **2 (9& 10)** : 7 – 10.

Dash G.(1998). Herbal use in Aquaculture. *Green Technology.* **2 (1& 2)** : 44-46.

Dash G., Parida, S. K. and Sahoo, S. N. (2009). Histopathological observations on carps infected with Argulus in the fresh water Bheries of West Bengal.*J.Appl.Biosci.* **35(2)**:159-162

Dash, G., Saha, S. and Nath, P.(2004). Prevalence and seasonal occurrence of intestinal helminth communities of Indian major carps in the freshwaters of West Bengal. *J. Environ. & Sociobiol* **1(1&2):** 21-28

Dash, G., Sahoo, S. N. and Dhali, M. (2009). Prevalence of Nematodes with special reference to Capillaria sp. (Chitwood et al., 1968) from major carps in India. Proccedings of the National seminar on aquaculture researches in India-status and strategies for future development. 26-27th March 2009. pp. 81-83.

Dash, G., Sahoo, S. N. and Parida, S. K. (2009). Haematological alternations in Indian Major Carps infected with *Argulus* in the Freshwater Bheries of West Bengal. Proccedings of the National seminar on aquaculture researches in India-status and strategies for future development. 26-27th March 2009. pp. 76-80.

Dash, G.,. and Parida, S. K., Sahoo, S. N and Roy, A. (2008). Haematological alternations carps infected with *Argulus* in the Freshwater Bheries of West Bengal. *Ind .J. Anim. Hlth.* **47(2)** : 77-82

Dash,G. Udgata,S.K. and Parida,S. (2008). Study of Helminth zoonotic parasites of carps in freshwater culture systems of West Bengal. *Ind. J. Animal Res.* **42(3)**:216-218.

Dash,G.(2000).Aquaculture needs fresh look in 21st century, *Indian J.Environ.& Ecoplan.,* **3(2)**:351-356.

Dash,G., Nath P.and Saha,S.(2004).Prevalence of nematodes with zoonotic potential in Indian Major Carps in the fresh waters of West Bengal. *J. Environ. & Sociobiol.* **1(1&2):** 55-60.

Dash,G., Parida, S.K., Sahoo, S.N. and Dhali, M.(2008) Consequences of Climate change verses sustainable developmentwith special reference to fisheries. *Aqua International* November pp. 26-29.

Dash,G.and Bandyopadhya,P.K.(2000).A Global Parasitological studies in EUS with special reference to India., *Environ. and Ecol.*, **18(1):**192-195.

Davidov, O.N., 1978. Growth development and fecundity of *Bothriocephalus gowkongensis* (Jeh, 1955), a cyprinid parasite. Hydrobiologicheski Zournal., **14:** 70–77 (in Russian, English translation: Hydrobiological J., **14:** 60–66, 1979).

Davis, H. S., Hoffman, G. L. and Surber, E. W. 1961. Notes on *Sanguinicola davisi* (Trematoda: Sanguinicolidae) in the gills of trout. *J. Parasitol.,* **47**: 512- 514.

Dehadrai, P. V. and Thakur, N. K. 1993. Air- breathing fish culture. In: Dhote, A. K. (ed.), *Fish culture*. National council of educational Research and Training, New Delhi. pp. 23- 44.

Denham, K. L. and Long, J. 1999. Occurrence of Gyrodactylus thymalli Zitnam, 1960 on grayling, *Thymallus thymallus* (L.) in England. *J. Fish Dis.* **22**: 247- 252.

Dezfuli, B. S. 1991. Histopathology in *Leuciscus cephalus* (Pisces : Cyprinidae) resulting from infection with Pomphorhynchus laevis (Acanthocephala). *Parasitologiya.* **33**: 137- 145.

Dick, T. A. and Choudhury, A. 1995. Cestoidea (Phylum : Platyhelminthes). In : Woo, P. T. K. (ed.) Fish diseases and disorders, vol. 1. Protozoan and Metazoan Infections. CAB International, London. Pp. 391-414.

Dugatkin, L.A., FitzGerald, G.J., and Lavoie, J., 1994, Juvenile three-spined sticklebacks avoid parasitized conspecifics, Environmental Biology of Fishes 39: 215-218

Dykova, I. & Lom, J., 1981. Fish Coccidia: critical notes on life cycles, classification and pathogenicity. *J. Fish Dis.,* 4: 487–505.

Erasmus, D. A. 1977. The host-parasite interface of trematodes. *Adv.In Parasitol.*, **15 :** 201-242.

Ergens, R. 1983. *Gyrodactylus* from Eurasian freshwater Salmonidae and Thymallidae, *Folia parasitologia*. **30**: 15- 26.

Eure, H. 1976. Seasonal abundance of *Neoechinorhynchus cylindratus* taken from large mouth bass (*Micropterus salmonoides*) in a heated reservoir. *Parasitol.* **73**: 355- 370.

Evans, W. A. 1974. Growth, mortality, and hematology of cutthroat trout experimentally infected with the blood fluke *Sanguinicola klamathensis*. *J. Wild. Dis.*, **10**: 341- 346.

Evlanov, I. A. and Kolokol'nikova, S. E. 1991. Population ecology of *Caryophyllaeus laticeps* (Cestoda, Caryophyllaeidae) in a bream population. *Zoologicheskiy.* **70**(10): 12-16.

Ewing, M.S. & Kocan, K.M., 1992. Invasion and development strategies of *Ichthyophthirius multifiliis*, a parasitic ciliate of fish. *Parasitology Today*, **8:** 204–208.

Ewing, M.S., Kocan, K.M. & Ewing, S.A., 1985. *Ichthyophthirius multifiliis* (Ciliophora) invasion of gill epithelium. *J. Protozool.* **32:** 305–310

Ferguson, H. W. 1989. Systemic Pathology of Fish. A Text and Atlas of Comparative Tissue Responses in Diseases of Teleost. Iowa State University Press/ Ames. U.S.A., pp. 1- 263.

Fernando, C. H. and Furtado, J. I. 1964. Helminth parasites of some Malaysian freshwater fishes. *Bull. Natural Museum Singapore*, **32**: 45- 71.

Fernando, C.H. & Furtado, J.I., 1963. A study of some helminth parasites of freshwater fishes in Ceylon. *Zeit. Parasitenkunde.*, **23**: 141- 163.

Frankland, H. M. T. 1955. The life history and bionomics of *Diclidophora denticulate* (Trematoda : Monogenea). *Parasitol.* **45**: 313- 351.

Fryer, G. 1968. The parasitic crustacea of African freshwater fishes; their biology and distribution. *J. Zool. Lond.* , **156** : 45-95.

Fryer, G., 1968a. The parasitic Crustacea of African freshwater fish: their biology and distribution. *J. Zool. London,,* **156:** 35–43.

Galli, P.,Crosa, G., Mariniello, L., Ortis, M. & Amelio, S. D 2001. Water Quality As A Determinant Of The Composition Of Fish Parasite Communities, Kluwer Academic Publishers. *Hydrobiologia* **452:** 173–179,.

GarciaI., F. Fujimoto, R.Y.; Martins. M.L., .MoraesI. F.R, 2009.Protozoan parasites of Xiphophorus spp. (Poeciliidae) and their relation with water characteristics, Arq. Bras. Med. Vet. Zootec. **vol.61 no.1** Belo Horizonte Feb. 2009.

Garnick, E., and Margolis, L., 1990, Influence of four species of helminth parasites on orientation of seaward migrating sockeye salmon (*Oncorhynchus nerka*) smolts, Canadian Journal of Fisheries and Aquatic Sciences **47:** 2380-2389.

Gault, N. F., Kilpatrick, D. J. and Stewart, M. T. 2002. Biological control of the fish louse in a rainbow trout fishery. *J. Fish Biol.* , **60** (1) : 226-237.

Ghag, A. 2004. Prevalence of helminth parasites with zoonotic potential in Indian Major Carps. M.F.Sc. Thesis submitted to West Bengal University of Animal and Fishery Sciences. pp:132

Ghosh, A. K., Dutta, N. C. and Laha, G. C. 1987. Observation on *dactylogyrid* trematodes of *catla catla* from Hooghly, West Bengal. *J. Inland Fish. Soc. India.* **19**(2): 53- 60.

Ghosh, D. 1990. " Wastewater- fed Aquaculture in the wetlands of Calcutta- an overview". *Wastewater fed Aquacult.* , pp : 50-56.

Giles, N., 1983, Behavioural effects of the parasite *Schistocephalus solidus* (Cestoda) on an intermediate host, the three-spined stickleback, *Gasterosteus aculeatus* L., Animal Behaviour **31:** 1192-1194.

Giles, N., 1987, Predation risk and reduced foraging activity in fish: experiments with parasitized and non-parasitized three-spined sticklebacks, *Gasterosteus aculeatus* L., Fish Biology **31:** 37-44.

GLOBEFISH Research Programme. 2001. FAO, **67**: 213 pp.

Godin, J.-G.J., and Sproul, C.D., 1988, Risk taking in parasitized sticklebacks *www.howfishbehave.ca* 8 under threat of predation: effects of energetic needs and food availability, Canadian Journal of Zoology **66:** 2360-2367.

Gopal Rao, K. 1988. Current fish disease problems in farmer's ponds of Andhrapradesh. *Asian Fish. Soc.Indian Branch. Newsletter.* , **1** : 3-6.

Gopalakrishna, V. 1964. Recent developments in the prevention and control of parasites of fish in Indian waters. *Proc. Zool. Soc. Bengal.* **17:** 95-100.

Gopalakrishnan, V. 1961. Observations on dropsy of Indian major carps and its experimental induction. *J. Sci. Ind. Res. (e),* **20**: 357- 358.

Gopalakrishnan, V. 1963. Controlling pests and diseases of cultured fishes. *Indian Livestock.* **1**(1): 51- 54.

Gopalakrishnan, V. 1968. Diseases and parasites in warm water ponds in Asia and the Far East, *FAO Fish Rep.,* **5** (44): 319- 343.

Gopalakrishnan, V. and Gupta, P. D. 1960. An eye disease which causes mortality of the fish *Catla catla* (Ham. & Buch.). *Curr. Sci.*, **29**: 240 pp.

Gupta, S. P. 1951. Three new trematodes of the family Hemiuridae Luhe, 1901 from freshwater fishes of U. P. *Ind. J. Helm.* **3**(1): 41- 54.

Gupta, S. P. 1953. Trematode parasite of freshwater fishes. *Ind. J. Helm.* **5**(1) : 3-12.

Gupta, S. P. 1961. Caryophyllaeids (Cestoda) from freshwater fishes of India. *Proc. Helm. Soc. Wash.* **28**: 38- 50.

Gupta, S. P. and Verma, S. L.1976. On a new acanthocephala *Acanthogyrus guptai n. sp.* from intestine of freshwater fish, *Labeo rohita* (Ham.) from Lucknow. *Riv. Parasitol.* **37**(2-3): 143- 150.

Hakalahti, T. and Valtonen, E. T. 2003. Population structure and recruitment of the ectoparasite *Argulus coregoni* Thorell (Crustacea: branchiura) on a fish farm. *Parasitology.* , **127** (1) : 79-85.

Hakalahti, T., Pasternak, A.F. and Valtonen, E.T. 2004. Seasonal dynamics of egg laying and egg-laying strategy of the ectoparasite *Argulus coregoni* (Crustacea: Branchiura). *Parasitology.* , **128** (6) : 655-660.

Halder, D. P., Das, M. K. and Sharma, B. K. 1983. Studies on the protozoan parasites from fishes. Four new species of the genera *Henneguya* Thelohan, 1892, *Thelohanellus Kudo*, 1933 and *Myxobolus butscheli*. 1882. *Arch Protestenk*, **127** (2): 283- 296.

Hamilton, W.D., and Zuk, M., 1982, Heritable true fitness and bright birds: a role for parasites? Science **218:** 384-386.

Harinasuta, C. 1984. Parasitic diseases of public health importance in southeast Asia-epidemiology, treatment and control. In : Ko, R. C. (ed). *Current Perspectives in Parasitic Diseases.* Department of Zoology and Medicine, University of Hong Kong, pp. 1-28.

Harrison, A. J., Gault, N. F. S. and Dick, J. T. A. 2006. Seasonal and vertical patterns of egg-laying by the freshwater fish louse *Argulus foliaceus* (Crustacea: Branchiura). *Dis. Aquat. Org.* , **68** (2) : 167-173.

Healy, G. R. 1970. Trematodes transmitted to man by fish, frogs and crustacea. *J. wildl. Dis.,* **6** : 255-261.

Heejeong Youn,(2009) Zoonotic Parasites, Korean J Parasitol. October; 47 (Supplement): S133–S141. Published online 2009 October 26. doi: 10.3347/kjp.2009.47.S.S133

Hibiya, T. 1982. *An Atlas of Fish Histology, Normal and Parasitological Features.* Kodansha Publ. Ltd. Japan., pp : 1-147.

Higashi, G. J. 1985. Food born parasites transmitted to man from fish and other aquatic foods. *Food Technol.,* **39** : 69.

Hindle, E. 1949. Notes on the treatment of fish infected with Argulus. *Proc. Zool. Soc. Lond.* , **119** : 79-81.

Hoffman, G. L. 1960. Synopsis of Strigeoidea (Trematoda) of fish and their life cycles. *Fishery Bull.,* **60**: 439- 469.

Hoffman, G. L. 1970. Control methods for snail-born zoonozes. *J. Wildl. Dis.*, **6**: 262- 265.

Hoffman, G. L. 1980. Asian tapeworm, *Bothriocephalus acheilognathi,* yamaguti, 1934, in North America. *Fish and Umwelt.* **8**: 69- 75.

Hoffman, G.L., 1980. Asian tapeworm, Bothriocephalus acheilognathi Yamaguti, 1934, in North America. Fisch und Umvelt, Special volume in honor of Prof. Dr H.-H. Reichenbach - Klinke 65th Birthday. **8:** 69–75.

Hoffman, R. W., Korting W., Fischer –Scherl, T. and schaufer, W. 1990. An outbreak of *Bucephalus polymorphus* in fish of the main river. *Angewandte parasitilogie* **31**: 95- 99.

Hoole, D., Bucke, D., Burgess, P. and Wellby, I. 2001. Infectious diseases- parasites. Diseases of carp and other cyprinid fishes. Fishing News Books, UK., pp. 63-122.

Hopkins, C. A. 1959. Seasonal variation in the incidence and development of the cestode *Proteocephalus filicollis* (Rud.1810) in *Gasterosteus aculeatus* (L. 1766). *Parasitol.* **49**: 529- 542.

Hora, S. L. and Pillay, T. V. R. 1962. Handbook on fish culture in the Indo Pacific Region. *FAO Fish Biol. Tech. Pap.,* **14** : 1-204.

Hora, S. L.1943. The fish louse Argulus foliaceous Linnaeus causing heavy mortality among carp fisheries of Bengal. Proc. Indian Sci. Cong. , **39 :** 66-67.

Hossain, A. M. and Dutta, H. M. 1991. Food deprivation induce differential changes in contents and micro- structures of digestive tract and appendages in bluegill fish, *Lepomis macrochirus,* comparative Biochemistry and Physiology. Part A. **100**: 769- 772.

Houde, A.E., and Torio, A.J., 1992, Effect of parasitic infection on male color pattern and female choice in guppies, Behavioural Ecology **3:** 346-351.

Hynes, H. B. N. 1950. The food of freshwater sticklebacks (*Gasterostaus aculeatus* and Pygosteus pungitius), with a review of methods used in studies of the food fishes. *J. Anim. Ecol.* **19**: 36- 58.

ICAR. 2006. Handbook of fisheries and aquaculture: pp. 354- 377, 511- 537.

J. Pan.2006. Cryptobia branchialis Nie from fish gills: ultrastructural evidence of ectocommensal function. *J. Fish dis.* , **3(5) :** 427-436

Jackson, P. B. N. 1978. Health and disease in intensive aquaculture. *J. S.A. Vet. Assoc.,* **49**: 57- 59.

Jafri, S.I.H. and Ahmed, S. S. 1991. A new record of ectoparasitic crustaceans (Branchiura: Argulidae) from major carps in Sindh, Pakistan. *Pak. J. Zool.* , **23** (1) : 11-13.

Jafri, S.I.H. and Ahmed, S. S. 1994. Some observations on mortality in major carps due to fish lice and their chemical control. *Pak. J. Zool.* , **6** (3) : 274 -276.

Jaideep-Kaur and Kaur, J. 1993. On some nematode parasites of freshwater fishes of Punjab. *Ind. J. Parasitol.,* **17**(2) : 193-198.

Jakobsen, P.J., Johnsen, G.H., and Larsson, P., 1988, Effects of predation risk and parasitism on the feeding ecology, habitat use, and abundance of lacustrine threespine stickleback (*Gasterosteus aculeatus*), *Canadian Fisheries and Sci.* 45: 426- 431.

Jayaram, K. C. 1977. Role of siluroid fishes in Aquaculture. *J. Ind Fish Soc. India.* **9**: 84- 91.

Jhingran, V. G. 1991. Fish and fisheries of India. 3rd Edition, Hindustan Publishing Corporation (India), Delhi. pp. 276- 291.

Johnsen, B. O. and Jensen, A. 1991. The *Gyrodactylus* story in Norway. *Aquaculture,*

Johnsen, B. O. and Jensen, A. J. 1986. Infestation of Atlantic Salmon, Salmo salar, by *Gyrodactylus salaris* in Norwegian rivers. *J. Fish Biol.,* **29**: 233- 241.

Johnsen, B. O. and Jensen, A. J. 1988. Introduction and establishment of *Gyrodactylus Salari* Malmberg 1957, on Atlantic Salmon (Salmo salar L.) fry and parr in River Vetsna, Norway. *J. Fish Dis.*, **11**: 35- 45.

Jubb, R. A. 1961. The freshwater eels (Anguilla spp.) of South Africa. An introduction to their identification and biology. *Annls. Cape Prov. Mus.*, **1**: 15- 48.

Kabata, Z. 1970. Crustacea as enemies of fishes. **In** Sneieszko, S.F. and Axelrod, H.R. (eds) *Diseases of Fishes*, Book 1. TFH Publ, New Jersey. , pp : 1 - 171

Kabata, Z. 1985. Injuries caused by crustacea. In Parasites and Diseases of Fish Cultured in Tropics. Taylor & Francis, London. , pp : 227- 270.

Kabata, Z. 1985. Parasites and Diseases of Fish Cultured in the Tropics. Taylor & Francis, London & Philadelphia. pp. 345.

Kabata, Z., 1970. Crustacea as enemies of fish. Book 1, Snieszko, S., Axelrod H.R. (ed.), *Diseases of Fish,* T.F.H. Publications, Jersey City, N.J.

Kabata. Z. 1985. Parasites and diseases of Fish Cultured in the Tropics. Taylor and Francis, London and Philadelphia.

Kennedy, C. R. 1969. Seasonal incidence and development of the cestoda *Caryophyllaeus laticeps* (Pallas) in the river Avon. *Parasitol*. **59**: 783-794.

Kennedy, C. R. 1995. Richness and diversity of macroparasite communities in tropical eels Anguilla reinhardtii in Queensland, Australia. *Parasitology.,* **111**: 233- 245.

Kennedy, C.E.J., Endler, J.A., Poynton, S.L., and McMinn, H., 1987, Parasite load predicts mate choice in guppies, Behavioural Ecology and Sociobiology **21:** 291-295.

Khalifa, K. A. 1986. Cestoda of freshwater farmed fishes in Iraq. *J. Wildl. Dis.*, **22**: 278.

Khalil, L. F. 1969. Studies on the helminth parasites of freshwater fishes of the Sudan. *J. Zool.* London, **158**: 143- 170.

Khalil, L. F. 1971. Check list of the helminth parasites of African freshwater fishes. Commonwealth Institute of Helminthology, St Albans. *Technical Communication No. 42. J. Sci.*, **29 (6):** 149- 153.

Khare, A. K., Ruhela, S. and Pandey, A. K. 2007. Histopathological manifestations in intestine, liver and kidney of *Clarias batrachus* carrying procamallanus infection. *Fishing chimes* **27**. (2) : 23- 26.

Khon, A. 1990. *Kritskyia moraveci n.g.*, n. Sp. (Monogenea, Dactylogyridae) from the urinary bladder and ureters of Rhamdia quelen (Quoy and Gaimard, 1824) (Pisces, Pimelodidae) in *Brazil Systematic Parasitol*. **17**: 81- 87.

Kim, K. H., Ahn, K. J. and Kim, C. S. 2001. Seasonal abundance of *Prosomicrocotyla gotoi* (Monogenea) and *Opecoelus sphaericus* (Digenea) from greenlings *Hexagrammos otakii* in a southern coastal area in Korea. *Aquacult.* **192**: 147- 153.

Kimura, S. 1970. Notes on the reproduction of water lice (*Argulus japonicus* Thiele). *Bull. Freshwater Fish. Res. Lab.* , **20** : 109 -126.

Knoff, M., Sao Clemente, S. C., Pinto, R. M. and Gomes, D. C. 2001. Digenea and Acanthocephala of elasmobranch fishes from the Southern coast of Brazil. *Mem. Inst. Oswaldo Croz., Rio de Janeiro.* **96**(8): 1095- 1101.

Koie, M. 1982. The redia, Cercaria and early stages of *Aporocotyle simplex* Odhner, 1900 (Sanguinicolidae) – A digenetic trematode which has a polychaete annelid as the only intermediate host. *Ophelia.,* **21:** 115- 145.

Korting, W. 1974. Bothriocephalosis of the carp. *Veterinar. Med. Nachr.*,**2**(74): 152- 158.

Krause, J., and Godin, J.- G.J., 1994, Influence of parasitism on the shoaling behaviour of banded killifish, *Fundulus diaphanus*, Canadian Journal of Zoology **72:** 1775-1779.

Krause, J., and Godin, J.-G.J., 1996, Influence of parasitism on shoal choice in the banded killifish (*Fundulus diaphanus*, Teleostei, Cyprinodontidae), Ethology 102: 40-49.

Kreutzmann, H. L. 1976. Haematological investigations on gill necrosis of the common carp *Cyprinus carpio. Z. Binnenfischerei DDR.* , **23** : 75 – 82.

Kruger, I., Van As, J. G. and Saayman, J.E. 1983. Observation on the occurrence of the fish louse *Argulus japonicus* Thiele, 1990 in Western Transvaal. *S. Afr. J. Zool.,* **18** : 408-410.

Kudo, R. R. 1993. A taxonomic consideration of myxosporidia. *Trans. Amer. Microsc. Soc.,* **52**: 195- 215.

Kuperman, B. I. and Davynov, V. G. 1982. The fine structure of glands in oncospheres, procercoids and plerocercoides of pseudophyllidea (cestoidean). *Int. J. Parasitol.* **12**: 135- 144.

Kurovskaya, L. A. 1992. Adaptive changes in the organism of carp infected with ectoparasites. *Gidrobiologicheskii-Zhurnal,* **28** (6) : 95-103.

Kurt Buchmann and T. Lindenstrøm, 2002. Interactions between monogenean parasites and their fish hosts, Int. *J. Parasitol.* **32:** 309-319.

Kurup, B. H. 1992. Effect of pH on disease development in freshwater environment. *Fishing Chimes*, **12**: 27- 33.

Lafferty, K.D., and Morris, A.K., 1996, Altered behaviour of parasitized killifish increases susceptibility to predation by bird final hosts, Ecology **77:** 1390-1397

Lalitha Kumari, P. S. 1969. Studies on parasitic protozoa (Myxosporidia) of freshwater fishes of Andhra Pradesh, *India. Riv. Parasitol.,* **30**: 153- 177.

Landsberg, J. H. 1989. Parasites and associated diseases of fish in warm water culture with special emphasis on intensification. In Shilo, M & Sarig, S. (ed) *Fish Culture in Warm water Systems: Problem and Trends.* CRC Press Inc, Boca Raton, Flo., pp : 195-252.

Landsberg, J.H., Blakesley, B.A. Reese, R.O., Mcrae, G. and Forstchen., P.R. 1998. Parasites of Fish as Indicators of Environmental Stress, Springer Link-Journal article (Environmental monitoring and Assesment **51:** 211-232,1998.)

Lom, J. and Molnar, K. 1983. *Myxobolus basilamellaris* in. *Myxosporea sp.* parasite of the gills of common carp (*Cyprinus carpio L.). Folia Parsitol*., **30** (1): 1- 3.

Lombard, G. L. 1968. A survey of fish diseases and parasites encountered in Transvaal. *Newsletter Limnol. Soc. S. Afr*., **11**: 23- 29.

Loot, G., Aulagnier, S., Lek, S., Thomas, F., and Guegan, J.F., 2002, Experimental demonstration of a behavioural modification in a cyprinid fish, *Rutilus rutilus* (L.), induced by a parasite, *Ligula intestinalis* (L.), Canadian Journal of Zoology **80:** 738-744.

Mackenzie, K. 1983. Parasites as biological tags in fish population studies. *Adv. Appl. Biol.* **7**: 251- 331.

Mackiewicz, J. S., Cosgrove, G. B. and Gude, W. D. 1972. Relationship of pathology to scolex morphology among Caryophyllaeid cestodes. *Z. f. Parasitenk*., **39**: 223- 246.

Madhavi, R., Umadevi, K. and Swarnakumari, V. G. M. 1997. Community structure of larval trematode fauna of the snail *Thiara tuberculata* from a freshwater stream at Visakhapatnam, Andhra Pradesh. *Current Sci*., **72 (8)** : 582-584.

Maiti, S. S.; Roy, R. N.; Haldar, D. P. 1937. Infection by a ciliate and a fluke on the fry of Indian major carps. *Curr. Sci.,* **56**(19) : 1029-1030.

Maity, K., Dash, G. and Das, N. (2003). Bhery a Potential Renewable source for waste disposal. *Environ. and Ecol.,* **21** (spl. Pub.) : 3 – 5.

Mallik, P. and Nandi, A. P. 2000. Nematode parasites of some fishes of Burdwan. *Trans. Zool. Soc. East India.* **4**(2): 28- 45.

Martin, M. 1932. On the morphology and classification of Argulus (Crustacea). *Proc. Zool. Soc. London.* , **10**: 771-806.

Mashego, S. N. 1982. A seasonal investigation of the helminth parasites of *Barbus species* in water bodies in Lebowa and Venda, South Africa. *Ph.D. thesis, University of the North, Sovenga, South Africa.* 191 pp.

McDonough, J. M. and Gleason, L. M. 1981. Histopathology in the rainbow darter, *Etheostoma caeruleum*, resulting from infections with the acanthocephalans *Pomphorhynchus bulbocolli* and *Acanthocephalus dirus*. *J. Parasitol.,* **67**: 403- 409.

McMinn, H. 1990. Effects of nematode parasite *Camallanus cotti* on sexual and non-sexual behaviour in the guppy (*Poecilia reticulata*). *American Zoologist*., 30 : 245-249.

Menezes, J., Ramos, M.A., Pereira, T.G. and Moreira da Silva, A. 1990. Rainbow trout culture failure in a small lake as a result of massive parasitosis related to careless fish introductions. *Aquaculture.* , **89**: 123-126.

Mikheev, V. N., Pasternak, A. F., Valtonen, E. T and Lankinen, Y. 2001. Spatial distribution and hatching of overwintered eggs of a fish ectoparasite, *Argulus coregoni* (Crustacea: Branchiura). *Dis. Aquat. Org.* , **46** (2) : 123-128.

Milinski, M., 1984, Parasites determine a predator's optimal feeding strategy, Behavioural Ecology and Sociobiology **15:** 35-37.

Milinski, M., 1985, Risk of predation of parasitized sticklebacks (*Gasterosteus aculeatus* L.) under competition for food, Behaviour **93:** 203-216.

Milinski, M., and Bakker, T.C.M., 1990, Female sticklebacks use male coloration in mate choice and hence avoid parasitized males, Nature **344:** 330-333.

Miller, D. M. and Dunagan, T. T. 1985. Functional morphology. In: Crompton, D. W. T., Nickol, B. B. (eds.), *Biology of the Acanthocephala*. Cambridge University Press, Cambridge. pp. 73- 123

Mishra, B. K 1991. Observations on the susceptibility of the hybrid, *Labeo rohita x Catla catla* to *Argulus siamensis* (Wilson) infection in composite culture ponds. *Proc. Nat. Symp. Fresh water Aquacult*. Bhubaneswar, Orissa. , pp : 84 - 88

Mishra, R. K. and Das, M. K. 1993. Ulceolariid ciliate *Trichodina reticulata* infesting gills of *Catla catla* in India. *J. Inland Fish. Soc. India.*, **25** (2): 54- 56.

Mitchell, A. J. and Hoffman, G. L. 1980. Important tapeworms of North American fresh water fishes. *U. S. Fish wild. Serv. Fish Dis. Leafl.*, **59**: 1- 18.

Mitra, K. 2002. Histopathology, An effective tool for assessing environmental impact on fish health. *Summer school on Methods of assessment of aquatic ecosytem for fish health care.* 18 July-16 August. , pp : 140-142.

Mizelle, J. D. and Klucka, A. R. 1953. Studies on monogenetic trematodes. XIV. Dactylogyridae from Wisconsin fishes. *Am. Midland Naturalist.,* **49:** 720- 733.

Mohan, C. V., Ramaiah, N and Shanbhogue, S. L. 1986. Effects of some therapeutics on fish ectoparasites. Environ. & Ecol. **4 (1)** : 98-100.

Mohan, C. V., Shankar, K. M and Ramesh, K. S. 1999. Mortalities of juvenile common carp, *Cyprinus carpio* associated with larval trematode infection – a case study. *J. Aquacult. Trop.,* **14 (2) :** 137-142.

Moller, H. 1978. the effects of salinity and temperature on the development and survival of fish parasites. *J. Fish Biol.*, **12 (4) :** 311-323.

Möller. H. 2006.The effects of salinity and temperature on the development and survival of fish parasites, *Journal of Fish Biology,* **12(4)** 311-323

Moravec, F. 1966a. On the systematics of genus *Paraquimperia* Baylis, 1934 (Nematoda : Quimperidae). *Helminthol.*, **7**: 353- 364.

Moravec, F. 1974a. The development of *Paracamallanus cyathopharynx* (Baylis, 1923) (Nematoda, Camallanidae). *Folia Parasitol.* (Praha), **21**: 333- 343.

Moravec, F. 1996. Some new data on the morphology of *Spinitectus inermis* (Nematoda: Cystidicolidae), a rare parasite of eels, as revealed by SEM. *Parasite.,* **3**: 169- 174.

Moravec, F., Boomker, J. and Taraschewski, H. 2000. *Paraquimperia Africana n. sp.* (Nematoda: Quimperiidae), a new intestinal parasite of the eel *Anguilla mossambica* Peters in South Arica. *J. Parasitol.*, **86**: 113- 117.

Mousavi, H. A. E. 2003. Parasites of ornamental fish in Iran. *Bull. Eur. Assoc. Fish Pathol.* , **23** (6) : 297-300.

Mukherjee, M. and Das, S. 2001. Artificial propagation of a Silurid fish, Pabda, *Ompok pabo* (H.). *Fishing Chimes*. **21**(1): 75- 79.

Museth, J., 2001, Effects of *Ligula intestinalis* on habitat use, predation risk and catchability in European minnows, *Fish Bio.* **59:** 1070-1080.

Musselius, V. A.; Strelkov, J. A.; Strelkov, V. A. and Strelkov, Yu, A. 1966. Parasites and diseases of the grass and silver carps in the fish farms of the USSR. *FAO Fisheries Reports,* **44**(5) : 353-360.

Muzall, P. M. 1995. Parasites of Pond-reared yellow perch from Michigan. *Progressive Fish-Culturist.*, **57**(2): 168- 172.

N. D. Thu ,A. Dalsgaard, L.T.T Loan and K.D.Murrell (2007) Korean *J. Parasitol.* **45(1):** 45-54, March .

Nandi, N. C. and Das, S. R. 1991. Argulosis causing Juveniles Mortality in some fishes in Kakdwip, West Bengal. *Indian J. Fish.* , **38** (2) : 132-133.

Natarajan, P. 1982. A new species of *Argulus muller* (Crustacea: Branchiura), with a note on the distribution of different species of Argulus in India. *Proc. Indian. Acad. Sci.* , **91** (4) : 375-380.

Nawa Y, Nakamura-Uchiyama F. (2004) An overview of gnathostomiasis inthe world. Southeast *Asian J Trop Med Public Health*; **35(Suppl** 1):87–91.

Nickol, B. B. 1995. Phylum Acanthocephala. **In** : Woo, P. T. K. (ed.), Fish diseases and Disorders, Vol. I. Protozoan and Metazoan Infections. CAB Int., London, pp. 447-478.

Nie, P. and Kennedy, C.R. 1991. Occurrence and seasonal dynamics of Pseudodactylogyrus anguillae (Yin and Sproston) (Monogenes) in eel, Anguilla anguilla (L.), in England. J. F. Biology, **39:** 897-900.

Niyogi, A. and Agarwal, S. M. 1989. Intestinal Pathology due to Caryophyllaeids parasitizing *Clarias batrachus* (Linn.) at Raipur. *Indian J. helminth.*, **41**: 112- 119.

Noga, E. J. 2000. A Text Book of Fish Disease: Diagnosis and Treatment. Iowa State University Press, Iowa, USA.

Northcott, S. J., Lyndon, A. R. and Campbell, A. D. 1997. An outbreak of freshwater fish lice, *Argulus foliaceus* L., seriously affecting a Scottish Stillwater fishery. *Fish. Manage. Ecol.* , **4** (1) : 73-75. Feb.

Oniye, S. J., Adebote, D. A. and Ayanda, O. I. 2004. Helminth parasites of *Clarias gariepinus* in Zaria, Nigeria. J. Aquatic Sci. **19** (2): 71- 76.

Overstreet, R. M. 1979. Crustacean health. Proc. of. 2[nd]Biennial Workshop. Tocas A and M Univ. College station. July 1979, TAMU-SG-79-114. , pp : 300-314.

Overstreet, R.M., Hawkins, W.E. & Fournie, J.W., 1984. The coccidian genus *Calyptospora* n.sp. and the family *Calyptosporidae* n. fam. (Apicomplexa), with members infecting primarily fishes. *J. Protozool.*, **31:** 332–339.

Owen, S.F., Barber, I., and Hart, P.J.B., 1993, Low level infection by eyefluke, *Diplostomum* spp. affects the vision of threespined sticklebacks, *Gasterosteus aculeatus*, *Fish Bio.* **42:** 803-806.

Padmavati, P. and Durgaprasad, M. K. 1998. An effective and economically feasible treatment of organophosphate pesticide and common salt to eradicate the fish ectoparasite, *Argulus japonicus* Thiele in carp culture ponds. *J. Environ. Biol.,* **19** (3) : 193-203.

Paperna, I . (1964). Host reaction on infestation of carp with Dactylogyrus vastator Nybelin, 1924 (Monogenea). Bamidgeh **16:** 129-141

Paperna, I. & Cross, R. H. M., 1985. Scanning electron microscopy of gamogony and sporogony stages of *Goussia cichlidarum* a coccidian parasite in the swimbladder of cichlid fishes. *Protistol.*, **21:** 473–479.

Paperna, I. & Zwerner, D.E., 1976. Parasites and diseases of striped bass, *Morone saxatilis* (Walbaum) from the lower Chesapeake bay. *J. Fish Biol.* **9:** 267–287.

Paperna, I. 1963a. Some observations on the biology and ecology of *Dactylogyrus vastator* in Israel. *Bamidgeh,* **15 :** 8-29.

Paperna, I. 1964. The metazoan parasite fauna of Israel inland water fishes. Bamidgeh *Bull. Fish. Cult. Israel.* **16**: 3- 66.

Paperna, I. 1991. Diseases caused by parasites in the aquaculture of warm water fish. Ann. Rev. Fish. Dis., **1 :** 155-194.

Paperna, I. 1995. Digenea (Phylum. Platyhelminthes). In : Woo, P.T.K. (ed.), *Fish Diseases and Disorders, Vol. I. Protozoan and Metazoan Infections.* CAB International, London. pp. 329-389.

Paperna, I. Hartley, A.H. & Cross, R.H.M., 1987. Ultrastructural studies on the plasmodium of *Myxidium giardi* (Myxosporea) and its attachment to the epithelium of the urinary bladder. *Int. J. Parasitol.* **17:** 813–819.

Paperna, I., 1980. *Amyloodinium ocellatum* (Brown, 1931) (Dinoflagellida) infestations in cultured marine fish at Eilat, Red Sea: epizootiology and pathology. *J. Fish Dis.*, **3:** 363–372.

Paria, T. and Konar, S. K. 1999. Management of fishponds and its relation to fish diseases in West Bengal, India. *Environ. & Ecol.*, **17**: 962- 970.

Parida, S.,Dash, G. and Udagata, S.K. (2008) Argulosis in carps with special reference to the water quality parameters of freshwater Bheries of West Bengal., *J.Inland. Fish. Soc. Ind. Spl. Issue* **40(1):** 162-166

Parker, R. 2002. Aquaculture Sciences, 2nd Edition. Delmar Pub., Thomson Learning, Inc., USA, pp. 331- 367.

Pathak, K. M. L. 1991. Fundamentals of parasitic zoonoses. Introduction. Kalyani Publishers, New Delhi. pp. 1- 17.

Pirus, R. I.; Makovetskaya, M. P. 1979. Dynamics of parasite fauna in young carp reared at a L'vov fish farm. *Rybn. Khoz.,* **29** : 81-84.

Plumb, J. A. 1994 Health maintenance of cultured fishes: Principal microbial diseases. CRC Publication, pp. 1- 239.

Poulin, R. and FitzGerald, G. J. 1988. Water temperature, vertical distribution and risk of ectoparasitism in juveniles stickle backs. *Can. J. Zool.*, **66** (9) : 2002-2005.

Poulin, R., and FitzGerald, G.J., 1989, Risk of parasitism and microhabitat selection in juvenile sticklebacks, Canadian Journal of Zoology **67:** 14-18.

Poulin, R., and FitzGerald, G.J., 1989, Shoaling as an anti-ectoparasite mechanism in juvenile sticklebacks (*Gasterosteus* spp.), *Behavioural Ecology and Sociobiology* **24:** 251-255.

Poulin, R., Rau, M.E., and Curtis, M.A., 1991, Infection of brook trout fry, *Salvelinus fontinalis*, by ectoparasitic copepods: the role of host behaviour and initial parasite load, Animal Behaviour **41:** 467-476.

Poulin, R., 1993, Age-dependent effects of parasites on anti-predator responses in two New Zealand freshwater fish, Oecologia 96: 431-438.

Poulin, R., 1994, Mate choice decisions by parasitized female upland bullies, *Gobiomorphus breviceps*, Proceedings of the Royal Society of London B 256: 183-187; López, S., 1999, Parasitized female guppies do not prefer showy males, Animal Behaviour 57: 1129-1134.

Poulin, R., 1999, Parasitism and shoal size in juvenile sticklebacks: conflicting pressures from different ectoparasites, Ethology 105: 959-968.

Prabhavathy, G. and Sreenivasan, A. 1976. Occurrence of *Argulus japonicus* in brood fishponds in Tamilnadu. *J. Inland Fish. Soc. India.* , **8** : 131-133.

Prasad, B. N. and Manjula, S. 1990. Ecological study of blue green algae in river Gomti. *Ind. J. Environ. Hlth.*, **22**: 151- 168.

Radabaugh, D.C., 1980, Changes in minnow, *Pimephales promelas* Rafinesque, schooling behaviour associated with infections of brain-encysted larvae of the fluke, *Ornithodiplostomum ptychocheilus*, *Fish Bio.* **16:** 621-628.

Rahman, M. R. and Jahan, M.S. 2002. Consequences of Larval Helminth Infecting Freshwater Gastropods. Bangladesh J. Zool. **30**(2), 101-114.

Ramakrishna, G. 1951. Notes on the Indian species of the genus *Argulus muller* (Crustacea : Copepoda) parasites on fishes. *Records of the Indian Museum.* , **49** (2) : 207-215.

Ranta, E., 1995, *Schistocephalus* infestation improves prey-size selection by three-spined sticklebacks, *Gasterosteus aculeatus*, Journal of Fish Biology **46:** 156-158.

Rao, K.H. & Ganapati, P.N., 1967. Observations on *Transversotrema patialensis* (Soparkar, 1924) (Trematoda) from Waltair, Andra Pradesh (India). *Parasitol.*, **57:** 661–664.

Ray, P. 1993. Aquaculture in Sunderban Delta, its perspective- an assessment, published by international Books and Periodical Supply Service., pp. 1- 197.

Ribelin, W. E and Walsh, A. H. 1975. The pathology of pesticide poisoning. In: *The Pathology of Fishes* (Ribelin, W. E. *et. al.,*) The University of Wisconsion Press. pp. 515- 557.

Robert J. Shields., Wilbur M. Tidd., 1968. Effect of Temperature on the Development of Larval and Transformed Females of Lernaea cyprinacea L. (Lernaeidae), JSTOR Crustaceana_supplement, No_1, pp 87-95.

Roberts, R. J. 1978. The pathophysiology and systemic pathology of fishes. In: A Textbook on Fish pathology. Ed. R. J. Roberts, Baillere Tindall, London.pp. 55- 91.

Roberts, R. J. 2001. Fish pathology, 3rd edition. WB SAUNDERS, Harcourt publishers ltd., pp. 457.

Roberts, R. J. 2001. The parasitology of teleost. In : Roberts, R. J. (ed.), Fish Pathology, W. B. Saunders, London, pp. 254-296.

Roberts, R. J., Macintosh, D. J., Tonguthai, K., Boonyaratpalin, S., Tayaputch, N., Phillips, M. J. and Millar, S. D. 1986. Field and laboratory investigation into ulcerative fish diseases in the Asia-Pacific region. Tech. Report of FAO Project TCP/RAS/4508. Bangkok, Thailand. 214 pp.

Rosenqvist, G., and Johansson, K., 1995, Male avoidance of parasitized females explained by direct benefits in a pipefish, Animal Behaviour **49:** 1039-1045.

Roy,A.,Maity,K., Dash,G .and Abraham,T.J.(2006). Microsporidiosis in Penaeus monodon in the Bheries of West Bengal ,India. *J. Aqua.* **14:**23-27

Rushton - Mellor, S. K. and Boxshall. 1994. The developmental sequence of *Argulus foliaceus* (Crustacea: Branchiura)., *J. Nat. Hist.*, **28** (4) : 763-785.

Safaa Z. Arafa, Mohammed M. El-Naggar and Samir A. El-Abbassy 2009. Mode of attachment and histopathological effects of *Macrogyrodactylus clarii*, a monogenean gill parasite of the catfish *Clarias gariepinus*, with a report on host respons, *Acta Parasitologica* **54:** 103-112

Saha, S. 2005. Study of helminth parasites of zoonotic potential with special emphasis on nematodes and trematodes in Indian Major Carps of West Bengal. *M.F.Sc. Thesis* submitted to West Bengal University of Animal and Fishery Sciences. pp. 134.

Sarig, S. 1971. The Prevention and Treatment of Diseases of Warm Water Fishes. *Diseases of Fishes.,* TFH Publ., New Jersy., Book 3., pp : 172

Sarkar, N. K. and Ghosh, S. 1990. Two new myxozoan parasites of the genus Thelohanellus kudo, 1993, (Myxosporea: Myxobolidae) from freshwater fishes of West Bengal, India. *New Agriculturist,* **1** (1): 35- 38.

Sasal, P., 2006, Nest guarding in a damselfish: evidence of a role for parasites, Journal of Fish Biology **68:** 1215-1221.

Schaeperclaus, W. (1954). Fischkrankheiten. Akademie- Verlag. Berlin W. Font, M. Collins, and S. Temple, Parasites as Indicators of Pollution ,PBRP(Lake Pontchartrain Basin Research Program) Research Highlight August 2007 .

Schaperclaus, W. 1933. Text book on pond culture: rearing and keeping of carps, trout and allied fishes (F. Hund Ed.). Fish Leaflet, Washington, **11**: 260.

Schaperclaus, W. 1991. Helminthiases. In : Schaperclaus, W., Kulow, H. and Schreckenoach. Fish Diseases. Vol. 2. Oxonoan Press, New Delhi, pp. 729-868.

Schmidt, G. D. 1985. Development and life cycles. In : Cromton, D.W.T. and Nickol, B.B. (eds.), *Biology of Acanthocephala*. Cambridge University Press, Cambridge. pp. 273- 305.

Scott, A. L. and Grizzle, J. M. 1979. Pathology of cyprinid fishes caused by *Bothriocephalus gowkongenesis* Yea, 1995 (Cestoda: Pseudophyllidea). *J. Fish Dis.,* **2:** 69- 73.

Seenappa, D. and Manohar, L. 1981. Five new species of Myxobolus (Myxosporea : Protozoa), parasitic in *Cirrhinus mrigala* (Hamilton) and *Labeo rohita* (Hamilton), with a note on a new host record for M. curmucae Seenappa and Mohahar , 1980, *J. Protozool.*, **28** (3): 358- 360.

Seppälä, O., Karvonen, A., and Valtonen, E.T., 2004, Parasite-induced change in host behaviour and susceptibility to predation in an eye fluke-fish interaction, Animal Behaviour **68:** 257-263.

Seppälä, O., Karvonen, A., and Valtonen, E.T., 2005, Impaired crypsis of fish infected with a trophically transmitted parasite, Animal Behaviour **70:** 895-900.

Seppälä, O., Karvonen, A., and Valtonen, E.T., 2005, Manipulation of fish host by eye flukes in relation to cataract formation and parasite infectivity, Animal Behaviour **70:** 889-894.

Shafir, A. and Van As, J. G. 1986. Lying, development and hatching of eggs of the fish ectoparasite *Argulus Japonicus* (Crustacea: Branchiura). *J. Zool.*, **210**: 401- 414.

Shella, F., Sivakumar, A. A. and Chandran, R. 2002. Infestation and prevalence of copepod parasite, *Argulus indicus* on some freshwater fishes. *Nature, Environ. Pollut. Technol.* , **1** (2) : 201-206.

Shimura, S. 1981. The larval development of *Argulus coregoni* Thorell (Crustacea: Branchiura). *J. Nat. Hist.* , **15** : 331-348.

Shimura, S. 1983a. Seasonal occurrence, sex ratio and site preference of *Argulus coregonii* Thorell (Crustacea: Branchiura) parasitic on cultured freshwater salmonids in Japan. *Parasitol.*, **86** (3): 537- 552.

Shirakashi, S., and Goater, C.P., 2001, Brainencysting parasites affect visually-mediated behaviours of fathead minnows, Écoscience **8:** 289-293.

Shirakashi, S., and Goater, C.P., 2002, Intensity-dependent alteration of minnow (*Pimephales promelas*) behaviour by a brain-encysting trematode, Journal of Parasitology **88:** 1071-1074.

Shirakashi, S., and Goater, C.P., 2005, Chronology of parasite-induced alteration of fish behaviour: effects of parasite maturation and host experience, Parasitology **130:** 177-183

Sindermann, C. J. 1990. Diseases caused by helminths and parasitic crustacea. **In** : Sindermann, C. J. (ed.), Principal Diseases of Marine Fish and Shellfish. Academic Press, New York, pp. 52-505.

Sineszko, S. F. and Axelrod, H. R. 1970. Diseases of Fishes. T.F.H. Publications, Inc. USA. , pp : 1-171.

Singh, H. S. and Sharma, M. 1992. *Wallagotrema indicus* n. sp., new monogenean from the fresh water sheat-fish *Wallago attu* Bloch and Schneider. *Functional and Development Morphology*. **2** (2): 85- 88.

Singhal, R. N., Jeet, S., and Davies, R. W. 1986. The relationships between changes in selected physicochemical properties of water and the occurrence of fish parasites in Haryana, India. *Trop.Ecol.* , **27** (1) : 1-9.

Singhal, R.N., Jeet, S. and Davies, R. W. 1986. Chemotherapy of six ectoparasitic diseases of cultured fish. *Aquaculture.* , **54** (3) : 165-171.

Smith, J. D. and McManus, D. P. 1989. The Parasitology and biochemistry of cestodes. Cambidge University Press, Cambridge. pp. 411- 493.

Smith, J. W. 1972. The blood flukes (Digenea : Sanguinicolidae and Spirorchidae) of cold blooded vertebrates and some comparison with *Schistosomes*. *Helminthol., Abs.* Ser. **A. 41** : 161-204.

Smith, J. W. 1972. The blood flukes (Digenea: Sanguinicolidae and Spirorchidae) of cold blooded vertebrates. Part I. A review of the literature published since 1971 and bibliography. *Helminthol. Abs.*, **66**: 254- 294.

Smith, R.S., and Kramer, D.L., 1987, Effects of a cestode (*Schistocephalus* sp.) on the response of ninespine sticklebacks (*Pungitius pungitius*) to aquatic hypoxia, *Canadian Zool.* **65:** 1862-1865.

Sneisko, S.F. 1974. The effect of environmental stress on out break of infectious diseases of fishes, J. Fish. Biol. **6**: 197-208

Snieszko, S. F., and Axelrod, H. R. 1970. Diseases of Fishes. T.F.H. Publications, Inc. USA, pp. 1- 171.

Snieszko, S. F.1983. Diseases of fishes: Research and Control. Fisheries. **8**, 20-22.

So, F. W. and Wittrock, D. D. 1982. Ultrastructure of the metacercarial cyst of Ornithodiplostomum ptychocheilus (Trematoda: Diplostomatidae) from the brain of fatheaded minnows. *Trans. Am. Microsc. Soc.*, **101**: 181- 185.

Sommerville, C. & Iqbal, N.A.M., 1991. The process of infection, migration, growth and development of *Sanguinicola inermis*, Plehn, 1905 (Digenea: Sanguinicolidae) in carp, *Cyprinus carpio* L. *J. Fish Dis.,* **14:** 211–219.

Soota, T. D. 1980. Collection and preservation of trematodes and cestodes. *Proc. Workshop Tech. Parasitol.* Zool. Surv. India. pp. 27-29.

Soulsby, E. J. L. 1982. Helminths. **In** : Soulsby, E. J. L. Helminths, Arthropods and Protozoa of domesticated animals. Seventh edition. The English Language Book Soc. and Bailliere Tindall. London, pp. 5-354.

Soulsby, E. J. L. 2005. Helminths, Arthropods and Protozoa of Domesticated Animals. Bailliere Tindall, 110 Greycoat Place, London SWIPISB. pp. v-ix.

Southwell, T. 1913. Notes from the Bengal Fisheries Laboratory. *Indian Museum. Rec. Ind. Mus.*, **9**(1): 79- 103.

Srivastava, C. B and Mukherjee, M. 1994. Parasitic infections in sewage-fed fishery of Sunderban delta. *Environ & Ecol.*, **12** (2): 441- 443.

Srivastava, C. B. 1980. Estimation of helminthic infections. *Proc. Workshop Tech. Parasitol. Zool. Surv. India.* pp. 29- 31.

Srivastava, C. V. 1975. Fish pathological studies in India: a brief review. In: Dr. B. S. Chauhan Comm., pp. 349- 358.

Stott, M.K., and Poulin, R., 1996, Parasites and parental care in male upland bullies (Eleotridae), Journal of Fish Biology **48:** 283-291.

Taraschewski, H. 2000. Host-parasite interactions in Acanthocephala: a morphological approach. *Adv. Parasitol.*, **46**: 1- 179.

Taylor, M.I., Turner, G.F., Robinson, R.L., and Stauffer, J.R. Jr, 1998, Sexual selection, parasites and bower height skew in a bower-building cichlid fish, Animal Behaviour **56:** 379-384.

Thilakaratne, I. D. S. I. P., Rajapaksha, G., Hewakopara, A., Rajapakse, R. P. V. J and Faizal, A. C. M. 2003. Parasitic infections in freshwater ornamental fish in Sri Lanka. *Dis. Aquat. Org.,* **54** (2) : 157-162.

Thomas, J. D. 1964. A comparison between the helminth burdens of male and female brown trout, *Salmo trutta* L, from a natural population in the River Teify, West Wales. *Parasitol.*, **54**: 263- 272.

Thompson R.C.A. (2000) Emerging parasite zoonoses. *Int. J. Parasitology* **30:**12–13.

Tierney, J.F., 1994, effects of *Schistocephalus solidus* (Cestoda) on the food intake and diet of the three-spined sticklebacks *Gasterosteus aculeatus*, *J. Fish Bio.* **44:** 731-735.

Tierney, J.F., Huntingford, F.A., and Crompton, D.W.T., 1993, The relationship between infectivity of *Schistocephalus solidus* (Cestoda) and anti-predator behaviour of its intermediate host, the three-spined stickleback, *Gasterosteus aculeatus*, Animal Behaviour **46:** 603- 605.

Tripathi, Y. R. 1952. Studies on parasites of Indian fishes. 1. Protozoa: Myxosporidia together with a checklist of parasitic protozoa described from Indian fishes. *Rec. Ind. Mus.,* **50**: 63-98.

Tripathi, Y. R. 1954. Studies on the parasites of Indian fishes. IV. Trematoda: Monogenea, Microcotylidae. *Rec. Ind. Mus.*, **52**: 231- 248.

Tripathi, Y. R. 1956. Studies on the parasites of Indian fishes. V. Acanthocephala. *Rec. Ind. Mus.* **54**(1-2): 61- 99.

Tripathi, Y. R. 1957. Monogenetic trematodes from fishes of India. *J Helminthol.,* **9**: 1- 149.

Tripathi, Y. R.1975. Studies on Branchiura from India fishes. *Dr. B.S Chauhan Commomeration Volume.* Zoological Society of India publication. , pp : 117-127.

Tweb, A. and Ahmed, A. 1981. Helminth infection in freshwater fishes of Bangladesh. *Fish Pathol.,* **15**(3/4) : 229-236.

Umadevi, K. 1996. *Studies on morphology, biology and surface ultrastructure of two species of haplochine trematodes* (Digenea). PhD Thesis, Andhra University, India, 214 p.

Vanotti, M.D., Tanzola, R. D. 2005. Relationships between the total parasite burden and some hematological parameters of Rhamdia sapo Val. (Pisces) under natural conditions. *Bull. Freshwat. Fish. Res. Lab., Tokyo.* , **20** (2) : 109-126.

W.H., 1952, The food of the cormorant in the Netherlands, Ardea 40: 1-63; Brassard, P., Rau, M.E., and Curtis, M.A., 1982, Parasite-induced susceptibility to predation in diplostomoiasis, Parasitology **85:** 495-501.

Walkey, M. 1967. The ecology of Neoechinorhynchus rutili (*Muller*). *J. Parasitol.*, **53**: 795-804.

Ward, A.J.W., Duff, A.J., Krause, J., and Barber, I., 2005, Shoaling behaviour of sticklebacks infected with the microsporidian parasite, *Glugea anomala*, Environmental Biology of Fishes **72:** 155-160.

Ward, A.J.W., Hoare, D.J., Couzin, I.D., Broom, M., and Krause, J., 2002, The effect of parasitism and body length on positioning within wild fish shoals, *Journal of Animal Ecology* **71:** 10-14.

Wedemeyer, G and Yasutake, W. T. 1977. Clinical methods for the environmental stress on fish health. *U.S. Fish. Wildlife Serv. Washington, D.C., U.S.A..* , **89** : 1-18.

Wedemeyer, G. A. 1996. *Physiology of Fish in Intensive Culture Systems.* New York: Chapman & Hall.

Wedemeyer, G. A., Meyer, F. P. and Smith, L. 1976. Diseases of Fishes: *Environmental Stress and Fish Diseases.* TFH. Publications, Neptune, NJ.

Wetzel, R. G. 1975. Limnology. W.B Saunders Co., Philadelphia, 743 pp.

WILLIAMS, H. & JONES, A. 1994. Parasitic worms of fish. London: Taylor & Francis.

Wilson, C. 1902. North American parasitic copepods of the family Argulidae, with a bibliography of the group and a systematic review of all known species. *Proc. US Nat. Mus.* , **34** : 635-783.

Woo, P.T.K., 2006. Fish diseases and disorders, Protozoan and metazoan infections, **(1):** 800

Yamaguti, S. 1958. System Helminthmum Vol-I, The Digenetic Trematodes of Vertebrates, Parts I & II. Interscience Publications, New York.

Yamaguti, S. 1959. Systema Helminthum, Vol. II, cestodes of vertebrates. *Interscience Publ. Inc.,* London, New York, 860 p.

Yamaguti, S. 1961. Systema Helminthum. Nematodes of vertebrate. 3. Part I and II, *Ibid.* P. 1-1261.

Yamaguti, S. 1963. Parasitic copepoda and branchiuran of fishes. Int. Sc. Publ., New York. , pp : 1-1104.

Yeh, I.S., 1955. On a tapeworm Bothriocephalus gowkongensis n. sp. (Cestoda: Bothriocephalidae) from freshwater fish in China. Acta Zool. Sinica, **7:** 69–74 (in Chinese, English summary.

Yoshikawa M, Nishimura F, Shiroi A.(2002) Behavioural observation of a sparganum mansoni surgically removed from the brain of a patient who developed a seizure [in Japanese]. *Clin. Parasitol*; **13:**129–31.

Zaman, Z. and Leong, T. S. 1987. Occurrence of the caryophyllid cestode *Lytocestus parvulus* Furtado, 1963 in *Clarias batrachus* (L.) in a tropical environment, Kedah, Malaysia. *J. Fish Biol.*, **31**: 591- 596.

Zdzitowiccki, K. and Pisano, E. 1996. New records on Digenea infection elasmobranch and teleost fish off Heard Island (Kerguelen, sub region, sub Antartic). *Archiv Fish Mar. Res.*, **43**: 265- 272.

Zdzitowiecki, K. 1989. New data on the morphology and distribution of two acanthocephalans, *Andracantha baylisi* (Zdzitowiecki, 1986) Comb. Nov. and *Corynosoma australe* Johnston, 1937. *Acta Parasitol. Polonica.*, **34**: 167- 172.

❑❑❑

INDEX

E

F

G

H

I

T

U

V

W

Y

Z